PHOTOSYNTHESIS

G. E. FOGG, Sc.D., F.R.S.

Professor of Marine Biology
University College of North Wales, Bangor

THE ENGLISH UNIVERSITIES PRESS LTD

ISBN 0 340 05081 0 Boards
ISBN 0 340 08373 5 Paperback

First printed 1968. Reprinted 1969 (with corrections), 1970 (with corrections)
Second edition 1972 Reprinted 1973

The English Universities Press Ltd
St Paul's House Warwick Lane London EC4P 4AH

Printed and bound in Great Britain by
Butler & Tanner Ltd, Frome and London

Preface

MANY books on photosynthesis have appeared in recent years. My justification for adding to their number is that few of those available at present are suited to the needs of students just beginning the study of the subject and most are of a narrowly specialist nature. It seems to me important that, in the excitement of unravelling the biophysics and biochemistry of the process, the integral part which photosynthesis plays in the life of plants and in the whole economy of the world should not be neglected. When food shortage threatens, the plant physiologist is still perhaps more useful than the molecular biologist. In writing this account I have therefore attempted to give as balanced and simple a review as possible of recent discoveries about the mechanism and to relate them to these wider aspects.

My gratitude must be expressed firstly to Professor J. E. Webb for his encouragement in the writing of this book and for his many helpful suggestions for its improvement. I am most grateful to my colleague Dr. D. O. Gray for reading the typescript and to Mr. A. D. Greenwood for his kindness in allowing me to choose from his many beautiful electron micrographs the seven which I reproduce as plates. My thanks are also due to Mrs. N. Last and Mr. R. Mitchell for their great help in preparing the typescript and illustrations respectively.

G. E. FOGG

Contents

List of Plates

List of Figures

1   *Background*

The nature of photosynthesis

A GREEN plant grows if given nothing more than air, water containing mineral salts in solution, and light. If there were such a person as a physical chemist who had no previous knowledge of plants he would find this phenomenon almost incredible.

Simple investigation shows that the elements of water constitute roughly 60 per cent of the solid material produced in the growth of a plant and that most of the remainder is derived from a gas, carbon dioxide, which is present to the extent of a small fraction of one per cent in the air. These unpromising raw materials are fully oxidized and hence of low potential chemical energy, whereas the product of growth, organic matter, is of high potential chemical energy, as shown by the fact that it will burn and liberate heat. This energy accumulated in the organic matter can only be derived from light. Such conversion of radiant energy to the potential chemical energy of stable organic compounds is only carried out by plants and certain bacteria; it does not happen in other kinds of living organisms and has not yet been imitated in the chemical laboratory.

It is not the conversion of radiant energy to chemical energy itself which is particularly remarkable in this process; this conversion happens, for example, when dyes fade in the sun and in the photographic process, but it then results in products which are extremely unstable, decomposing immediately with the liberation of their energy as heat so that the end result is a loss of potential chemical energy. The astonishing thing about plants is that they avoid such back-reactions and fix light energy in the form of potential chemical energy in compounds which do not break down spontaneously and which can be stored until required later on for the carrying out of life processes. It is this which is the essence of *photosynthesis*.

This view of photosynthesis obviously depends on fairly advanced chemical and physical concepts and could not have been conceived until the necessary background of knowledge had been provided. By the beginning of the nineteenth century analytical procedures of sufficient precision had been devised and it was established with their aid that green plants are able to produce organic matter in the light from carbon dioxide and water, a quantity of oxygen roughly equivalent to the carbon dioxide absorbed being given off. This may be summed up in the equation:

$$1) \qquad\qquad CO_2 + H_2O \xrightarrow{\text{light}} (CH_2O) + O_2$$

in which the organic matter is denoted by the approximate empirical formula (CH_2O). The general picture was completed when it was realized that the process involved the transformation of radiant energy into chemical energy. This conclusion follows from the law of the conservation of energy and it was the propounder of this law himself, the German surgeon Mayer, who in 1845 pointed out its implications in the case of photosynthesis. This established the rôle of plants in nature as the primary producers of organic matter on which animals, fungi and non-photosynthetic bacteria depend for the chemical energy necessary for the maintenance of life.

The rôle of chlorophyll

Photosynthesis occurs only in the green parts of plants. This, together with the necessity for light, was shown in 1779 by the Dutch physician Ingen-Housz as the main outcome of 500 experiments carried out during the course of a summer's vacation in England (Plate I). The green colour is confined to small sub-cellular organelles known as *chloroplasts*, first described in 1837, and these were in due course shown to be the site of photosynthesis. Starch, which is the chief product of photosynthesis in mature leaves, was shown by the German plant physiologist, Sachs, in 1862, to accumulate inside chloroplasts during illumination. Then in 1883, by ingenious use of bacteria which migrate towards concentrations of oxygen another German, Engelmann, demonstrated that this gas is liberated only from chloroplasts and not from the colourless parts of the protoplasm (Plate II*a*).

It was not obvious to all nineteenth-century plant physiologists that light must be absorbed in order to have a chemical effect and that therefore a pigment must be involved, but the invariable association of the capacity for photosynthesis with the presence of green chloroplasts pointed to the importance of the pigments. The isolation and characterization of the chloroplast pigments, the green chlorophylls *a* and *b*, and the two yellow carotenoids, carotene and xanthophyll (now known as lutein), were briefly reported by the English physicist Stokes in 1864. The chromatographic method of separation that later was to prove such a powerful means of studying not only the pigments themselves but also the colourless initial products of photosynthesis was described by the Russian botanist Tswett in 1906. It was not used, however, in the classical studies of the German chemist, Willstätter, and his collaborators (1913–1918), who preferred more orthodox methods. They showed that whereas the same pigments are present in all higher plants, seaweeds, although they always have chlorophyll *a*, may or

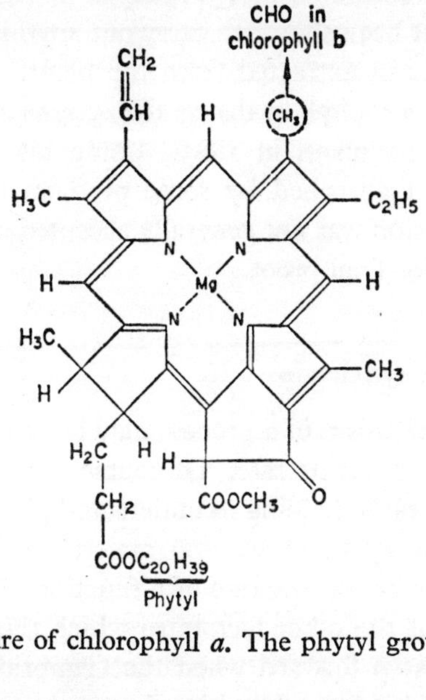

Figure 1 The structure of chlorophyll *a*. The phytyl group is a long unbranched chain.

may not have other chlorophylls. They may also have characteristic xanthophylls, such as fucoxanthin in the brown seaweeds.

The amounts of chlorophylls *a* and *b* were found to remain constant during short periods of illumination, thus indicating that their rôle is catalytic. Because the chlorophyll molecule contains magnesium, Willstätter and Stoll in 1918 suggested that its function is to form a compound with carbon dioxide, in a manner analogous to the well-known reaction in organic chemistry in which carbon dioxide is added onto an organo-magnesium compound, and that this becomes altered in the light to a formaldehyde–peroxide compound which then splits to give formaldehyde (CH_2O), from which sugars may be formed by condensation. This was an elaboration of the formaldehyde hypothesis first put forward in 1870 by Baeyer. It was an exercise

in speculative biochemistry for which neither Willstätter and Stoll nor subsequent workers could find direct evidence and it is now known to be in-
correct. However, the work of Willstätter and Stoll was immensely important
even though uncritical acceptance of ideas to which they gave their authority
has in several instances slowed down the progress of research in photosynthesis.

The chemical structure of chlorophyll *a* (see Fig. 1) was not finally settled
until 1940, when the German chemist Fischer provided conclusive proof by
synthesis. It is clear that this substance is the central one in photosynthesis
but knowledge of its structure has not yet helped in suggesting the exact rôle
which it plays nor has it been shown to carry out anything resembling photosynthesis once it has been extracted from the plant. That the chloroplast
pigments other than chlorophyll *a*, that is the *accessory pigments*, may play a
part was shown by Engelmann in 1883. Using his bacterial method, he
demonstrated that light absorbed by some of them is effective in photosynthesis. This conclusion was not generally accepted at the time and had to
await the modern era for final proof.

The kinetics of photosynthesis

Useful clues to the mechanism of a process may be obtained by studying the
effects of different factors on its rate. Of course, *kinetic* studies of this sort
are also of general interest in helping to understand plant growth. At the end
of the nineteenth century there was no general realization among plant
physiologists that the effect on the rate of a function of change in one factor
depends on the level of the other factors to which the plant is exposed. It
therefore was a great step forward when the Cambridge plant physiologist
Blackman put forward his law of limiting factors in 1905. This stated that
when a process is affected by several factors its rate is determined by the
factor which is in shortest supply. It was later shown that this is not always
exactly true but it still remains a good general guide to plant behaviour.
Applied to photosynthesis it led to some interesting conclusions. Blackman
and Matthaei found that whereas increase in temperature in bright light leads
to an increased rate of photosynthesis, there is little effect from change in
temperature if the light intensity is low. Now, ordinary chemical reactions
usually double or treble in rate for a 10°C rise in temperature, but reactions
utilizing light energy, that is *photochemical* reactions, are temperature independent provided that they are not limited by supply of reactants. Blackman concluded from this that photosynthesis does not consist of just a photochemical reaction, but that one or more purely chemical ("dark") reactions

must be involved as well. At low light intensities the rate of photosynthesis is limited by a photochemical reaction, but at high light intensities the "dark" reactions become limiting and light energy absorbed over and above that required to saturate their requirement for the products of the photochemical reaction is wasted. This situation is shown in Fig. 2, from which it will be seen that at low intensities the rate of photosynthesis is proportional to light intensity, whereas at higher intensities the rate reaches a plateau and is

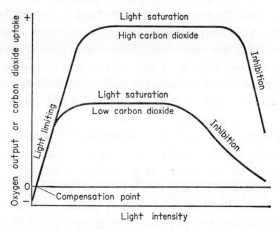

Figure 2 The relationship of rate of photosynthesis to light intensity as shown by a single leaf or a suspension of algal cells. The two curves are for different carbon dioxide concentrations.

independent of intensity. Supply of carbon dioxide at a higher concentration, by enabling the "dark" reactions to go faster, increases the intensity at which this plateau is attained.

The idea that "dark" reactions are involved in photosynthesis received support from the finding of the German biochemist Warburg, reported in 1919, that the photosynthetic yield from a given amount of light is greater if it is given in short intense flashes separated by dark periods than if it is given continuously. Here, evidently, the rate-limiting "dark" reactions are able to catch up with the light reaction during the intervals of darkness so that a greater overall rate of photosynthesis is achieved. In this work Warburg introduced both his manometric method of measuring gas exchanges of living cells, and the use of the microscopic unicellular alga *Chlorella* as an experimental subject in which the functioning of a green plant could be studied in the simplest terms. Both of these techniques were to play an immensely important part in future studies on photosynthesis.

Warburg then turned his attention to the efficiency of photosynthesis. The energy of radiation is distributed in packets, *quanta*, which can only be absorbed as complete entities, the energy content of red quanta being such that at least three are required for the reduction of a molecule of carbon dioxide to the level of carbohydrate. Warburg and Negelein, in 1922, determined the actual quantum requirement as being four, implying a remarkably high efficiency, and independent of the wavelength of the light used. Although straightforward in principle the determination of the quantum efficiency of photosynthesis has subsequently given rise to much controversy —often acrimonious. The major trouble perhaps has arisen because at low light intensities, at which determinations of quantum efficiencies must . necessarily be made, the corrections needed to allow for respiration and other processes are relatively large and vary in a not altogether predictable manner with the history of the cells. It is now generally agreed that at least eight quanta are required for the reduction of one molecule of carbon dioxide to the carbohydrate level.

Further investigation of the effects of flashing light reported in 1932 by Emerson and Arnold in America showed that the number of carbon dioxide molecules reduced by an intense flash is never more than one per 2500 chlorophyll molecules. This suggested a "photosynthetic unit", a block of chlorophyll molecules in which the energy of quanta absorbed by any four or more could be funnelled into the reduction of one carbon dioxide molecule. Studies of kinetics thus gave considerable insight into the mechanism, but in the absence of information about the chemical nature of the reactants and intermediates the picture was incomplete and unsatisfying.

Comparative biochemistry

In higher plants photosynthesis may appear to be a stereotyped process but among the lower plants distinct variations occur. Photosynthetic bacteria of various types are common in situations where organic decomposition is going on with the production of hydrogen sulphide and other reduced compounds. These bacteria do not possess chlorophyll *a* but instead have similar substances, bacteriochlorophyll in purple bacteria and chlorobium chlorophyll in the green bacteria. The purple sulphur bacteria, first described in 1882 by Engelmann in Holland and in 1888 by Winogradski in Russia, are light dependent but do not evolve oxygen. For this reason there was controversy about their nutrition until the 1930s when the studies of van Niel, a Dutchman working in America, cleared up the confusion. He showed that both purple and green sulphur bacteria are capable of growing in a purely inorganic medium providing that light and hydrogen sulphide are available.

Hydrogen sulphide is oxidized and carbon dioxide assimilated by green sulphur bacteria in accordance with the equation:

$$2) \qquad 2H_2S + CO_2 \xrightarrow{\text{light}} (CH_2O) + H_2O + 2S$$

Purple sulphur bacteria (*Chromatium*) and some species of green sulphur bacteria (*Chlorobium*) may oxidize the sulphide to sulphate and bacteria of both genera may utilize molecular hydrogen or thiosulphate instead of hydrogen sulphide. The essential feature is that this form of photosynthesis, in which no trace of oxygen is evolved, requires a supply of a substance capable of giving up hydrogen, that is, a *hydrogen donor*. Purple sulphur and non-sulphur bacteria (of which *Rhodospirillum rubrum* is the most studied) may use organic as well as inorganic substances as hydrogen donors. A connexion between these forms of photosynthesis and that of green plants was provided in 1939 by Gaffron in Warburg's laboratory when he showed that the alga *Scenedesmus* becomes capable of a type of photosynthesis, called *photoreduction*, similar to that of the purple sulphur bacteria, after a period of adaptation under anaerobic conditions, the gas exchanges then being in accordance with the equation:

$$3) \qquad CO_2 + 2H_2 \xrightarrow{\text{light}} (CH_2O) + H_2O$$

van Niel pointed out that a common pattern may be seen in these various types of photosynthesis if it is assumed that the basic process is the splitting of the water molecule:

$$4) \qquad H_2O \xrightarrow{\text{light}} [H] + [OH]$$

the [H] fragment, which is not envisaged as being free hydrogen, being used for the reduction of carbon dioxide,

$$5) \qquad 4[H] + CO_2 \rightarrow (CH_2O) + H_2O$$

In the photosynthetic bacteria and adapted algae the [OH] was supposed to be removed by combination with hydrogen from the hydrogen donor to give water, e.g.:

$$6) \qquad 4[OH] + 2H_2S \rightarrow 4H_2O + 2S$$

The higher plants were supposed to have evolved a method of disposal which eliminates oxygen in its molecular form:

$$7) \qquad 4[OH] \rightarrow 2H_2O + O_2$$

It will be noticed that this is a different picture of photosynthesis from that presented, for example, in the theory of Willstätter and Stoll in which the oxygen is supposed to originate by decomposition of the carbon dioxide. That van Niel's view on the origin of the oxygen is essentially the correct one has been confirmed by using the isotope ^{18}O as a tracer, but his interpretation of bacterial photosynthesis has required modification in the light of recent research (see p. 54).

Identification of intermediates

It is an obvious approach to determining the nature of the first products to compare analyses of plants that have been kept for some time in the dark with those of similar plants after illumination for a short period. Since it appeared that the product contained the elements C, H and O in the same proportions as in a carbohydrate, attention was first mainly concentrated on this class of substance in such experiments. The classical methods of chemical analysis used were mostly cumbrous and insensitive and now that we know that the concentration of a photosynthetic intermediate may wax and wane within a matter of seconds following change in illumination, it is not surprising that no definite results emerged from this type of experiment. Reviewing this phase of investigation in 1949, Smith could only conclude that diverse substances—various carbohydrates, proteins or acids—may arise simultaneously as a result of photosynthetic activity.

The situation changed altogether when the American workers Calvin and Benson in 1948 combined the techniques of paper chromatography and the use of radiocarbon, ^{14}C, as a tracer. ^{14}C behaves chemically in the same way as the common stable isotope, ^{12}C, except that, being heavier, it causes the molecules containing it to react slightly more slowly. A plant thus assimilates $^{14}CO_2$ as it would $^{12}CO_2$, the molecules into which it is incorporated becoming radioactive. By means of chromatography minute amounts of intermediates present after a few seconds of illumination can be separated from the mixture obtained in an extract of the plant and then identified by means of their radioactivity (Plate II*b*). Phosphoglyceric acid was quickly recognized as being the first stable product of photosynthetic carbon fixation and by 1954 the cycle, which is still generally accepted, whereby ribulose diphosphate gives rise by addition of carbon dioxide to two molecules of phosphoglyceric acid and is in turn regenerated from this substance (see p. 65), was put forward. In the cycle the potential chemical energy of adenosine triphosphate (ATP), and a hydrogen donor, reduced nicotinamide adenine dinucleotide phosphate (NAPDH$_2$, also known as coenzyme II or triphosphopyridine nucleotide TPNH$_2$) are utilized for the reduction of carbon dioxide to the carbohydrate level. In 1955 Racker demonstrated the synthesis of carbohydrate from bicarbonate in a multienzyme system derived from rabbit muscle, yeast and spinach leaves. The mixture contained beside the enzymes concerned in the cycle, ATP and molecular hydrogen (which in the presence of an appropriate enzyme system yielded the hydrogen donor) but no chlorophyll. It thus seems that the photosynthetic fixation of carbon dioxide is carried out by a mechanism which does not involve photochemical reactions directly and which, in fact, has no features that cannot be found in non-photosynthetic organisms.

The reactions of isolated chloroplasts

These results were mainly accomplished with intact plant tissues. Obviously it would be a great help in investigation if photosynthesis could be separated from the multifarious other processes which go on in a living cell. That the biochemical mechanism of fermentation was largely worked out by 1940 can be attributed mainly to the circumstance that it is possible to obtain from yeast a cell-free juice which is still capable of fermentation. For a long time it seemed that photosynthesis could not be isolated in this way and up to 1940 it was generally believed that chloroplasts needed the co-operation of some other protoplasmic component in order to carry out photosynthesis.

The first successful step towards the goal of isolating the photosynthetic machinery in a working condition from the rest of the cell was reported in 1939 by the Cambridge biochemist, Hill. He obtained chloroplast suspensions from ground-up leaves of chick-weed (*Stellaria media*) and other plants, which when illuminated evolved oxygen provided that a suitable hydrogen-accepting substance, such as ferric oxalate, was present. Only certain species yield active preparations. Willstätter and Stoll among others had also made studies with chloroplast preparations and might well have obtained positive results had they chosen, say, spinach instead of *Pelargonium* leaves. Subsequently other hydrogen acceptors for the Hill reaction were found; these included benzoquinone and the naturally occurring NADP which has just been mentioned as implicated in the carbon fixation cycle, but not carbon dioxide itself. The Hill reaction and photosynthesis seem to be related since they are affected in a similar way by a variety of treatments. These isolated chloroplasts thus seem to contain part of the photosynthetic mechanism, that which, on the basis of van Niel's ideas, is responsible for the splitting of water:

8)
$$4H_2O \xrightarrow{\text{light}} 4[H] + 4[OH]$$

9)
$$4[OH] \longrightarrow 2H_2O + O_2$$

10)
$$4[H] + A \longrightarrow 2AH_2$$

where A is the hydrogen acceptor (the Hill reagent).

About this time the rôle of organic phosphates such as ATP in the energy transactions of living organisms had been discovered and many attempts were made to find out whether these substances were produced at the expense of light energy. The evidence remained inconclusive until Arnon and Whatley in California in 1954 showed that isolated chloroplasts are able to produce ATP from adenosine diphosphate (ADP) and inorganic phosphate (P_i) in the light:

11) $$ADP + P_i \xrightarrow{\text{light}} ATP$$

In the same year a similar reaction was reported as occurring in photosynthetic bacteria by Frenkel. This reaction, in which no oxygen is produced, Arnon termed *cyclic photophosphorylation*. In it there is a conversion of light energy into the potential chemical energy of the third phosphate bond in ATP. The possibility that the ATP was being produced by ordinary respiratory processes from a product of the photochemical reaction was ruled out by the fact that photophosphorylation occurs in chloroplast preparations which are free from mitochondria (the organelles in which oxidative phosphorylation, the ATP-producing reaction of respiration, occurs) and which show no respiratory activity. Moreover, photophosphorylation can take place in the absence of oxygen provided that certain co-factors are provided. Arnon's chloroplast preparations were also capable of carrying out the Hill reaction and he found evidence that in the presence of a suitable Hill reagent photophosphorylation was coupled with its reduction:

12) $$2ADP + 2P_i + 2NADP + 2H_2O \xrightarrow{\text{light}} 2ATP + 2NADPH_2 + O_2$$

This was termed *non-cyclic* or *synthetic photophosphorylation*. It may be noted that chloroplasts lose soluble components in the course of isolation so that in order to demonstrate non-cyclic photophosphorylation it is necessary to add a chloroplast extract or factors such as ferredoxin and NADP-reductase back to the washed chloroplast suspension. The Hill reaction seems to be an artificial variant of non-cyclic photophosphorylation in which the phosphorylating steps have become uncoupled and in which the natural hydrogen acceptor is replaced by an unnatural one. The products of non-cyclic photophosphorylation are just those which are required to drive the carbon fixation cycle discovered by Calvin and Benson and the reaction is often spoken of as generating "assimilatory power".

Later it was found that with special precautions chloroplasts retaining a capacity for complete photosynthesis of carbohydrates from carbon dioxide may be obtained (see Plates VI and VII). In an elegant experiment reported in 1958 Trebst, Tsujimoto and Arnon showed that in such chloroplasts assimilatory power and carbon dioxide assimilation can be separated. This was done by first illuminating them in the presence of ADP, inorganic phosphate and NADP, but in the absence of carbon dioxide, to produce ATP and $NADPH_2$. Following this, the insoluble green residue was removed and the remaining solution incubated in darkness with ^{14}C-labelled carbon dioxide whereupon the usual products of photosynthetic carbon fixation, including sugar phosphates, became labelled. Isolated chloroplasts are now widely used in research on the mechanism of photosynthesis. There is still some uncertainty as to how far the reactions which they carry out occur in the intact plant but undoubtedly they provide one of the best means of investigation available.

Photosynthesis in the life of the plant

Before we consider details of mechanism any further, a general comment should be made. Many of the men who pioneered in the investigation of photosynthesis were primarily chemists and today research on its mechanism is largely a matter for specialist biochemists and biophysicists. Quite understandably these are inclined to regard all parts of the plant except the chloroplasts as so much extraneous matter that has to be cleared away before work can begin. On the other hand the botanist is tempted to look on photosynthesis as a process the results of which he can take for granted, putting aside the mechanism as something beyond his comprehension. Such tendencies to regard photosynthesis as separate and distinct from the rest of a plant's activities were perhaps reinforced by theories such as that of Willstätter and Stoll, mentioned above, in which the mechanism is visualized as being of a different nature from the rest of metabolism, the only point of contact between the two being in the form of a definite end-product, sugar. However, as we have already seen, the substances which most justifiably may be called the primary products of photosynthesis are the high-energy phosphate, ATP, and the hydrogen donor, $NADPH_2$. These are known to be involved in a wide variety of biochemical reactions so that, in principle at least, photosynthesis may directly intermesh with and influence almost any of the life processes of the plant. The discoveries of Hill, Arnon, and Benson and Calvin have definitely disposed of the old idea of photosynthesis as a biochemically isolated process. There is interaction in both directions; not only do the activities and behaviour of a green plant normally depend directly on photosynthesis but in turn there is a feed back, the rate and ultimate products of photosynthesis being influenced by the growth, development and physiological state of the plant body.

Further Reading

Fogg, G. E. (1970) *The Growth of Plants*. 2nd edition, Penguin Books, Harmonds-
    worth.
Gaffron, H. (1960) Energy storage: photosynthesis. In *Plant Physiology*, edited by
    F. C. Steward, vol. I B, pp. 3–277. Academic Press, New York and London.
Hill, R., and Whittingham, C. P. (1957) *Photosynthesis*. 2nd edition, Methuen,
    London.
Loomis, W. E. (1960) Historical introduction. In *Encyclopedia of Plant Physiology*,
    edited by W. Ruhland, vol. V (1), pp. 85–114. Springer-Verlag, Berlin.
Rabinowitch, E. I. (1945, 1951, 1956) *Photosynthesis and Related Processes*. Vols. I
    II (1) and II (2). Interscience, New York and London.
Rabinowitch, E. I., and Govindjee (1969) *Photosynthesis*. John Wiley & Sons Inc.,
    New York and London.
San Pietro, A., Greer, F. A., and Army, T. J. (Editors) (1967) *Harvesting the Sun:
    Photosynthesis in Plant Life*. Academic Press, London and New York.
Van Niel, C. B. (1962) The present status of the comparative study of photosyn-
    thesis. *Ann. Rev. Plant Physiol.*, **13**, 1–26.

2 *The supply of light, carbon dioxide and water*

PHOTOSYNTHESIS by green plants and algae normally depends on sunlight and consumes water, as the source of the hydrogen used in reduction, and carbon dioxide, as the main acceptor for this hydrogen. The supplies of light and these two principal raw materials deserve attention because, not only are they the chief factors limiting the rate of the process under natural conditions, but we may suppose that their availability has in the course of evolution determined the general structure and functioning of the photosynthetic apparatus.

Sunlight

The radiation reaching the surface of the Earth from the sun has a maximum intensity of about $1 \cdot 6$ cal per cm^2 per minute, corresponding to a visible irradiance of about 100,000 lux or metre-candles. Most of this energy is concentrated in wavelengths between 600 and 900 milli-microns (1 milli-micron or $m\mu = 1$ nm $= 10^{-9}$ m $= 10$ Ångstrom units), that is to say, in yellow and red light and near infra-red radiation, but the spectrum extends well into the ultra-violet on the short wavelength side and through the infra-red to radio frequencies in the longer wavelengths. The intensity, and to a lesser extent, the spectral composition, of solar radiation varies, of course, with time of day, weather, season and geographical position. Plants must make the best use they can of this mixed and variable radiation for the energy they need for photosynthesis.

At the equator the maximum solar radiation which can be received on a horizontal surface varies between 780 and 900 cal per cm^2 per day according to the season. At the latitude of London ($51°\ 30'$ N) the value is about 980

cal per cm^2 per day in summer, being higher than that at the equator because
the day is longer, and about 170 cal per cm^2 per day in the shortest days of
winter. On Spitsbergen (80° N) the value reaches 1050 cal per cm^2 per day at
midsummer when the sun is above the horizon for twenty-four hours and is
zero for over four months in the winter. The radiation actually received by a
particular surface at any given time consists of direct sunlight and indirect
solar radiation from the sky. The former varies with the angular height of the
sun, which depends on time of day, season and latitude. The orientation of
the receiving surface is, of course, important; little increase in incident solar
radiation has been found for orientations other than horizontal in midsummer
but at midwinter in northern latitudes a south-facing surface tilted at an
appropriate angle receives distinctly more radiation in the course of a day
than either a horizontal or a vertical surface. Added to these complications
are those of weather. An average cloudy sky reduces the midday light intensity to between 15 and 5 per cent of that when the sky is cloudless. The
average annual duration of sunshine varies from below 800 hours, for example
in such persistently foggy places as the Aleutian Islands, to over 4000 hours
in the Sahara and Californian deserts, which have over 90 per cent of the
maximum possible. Southern England has an average of about 1600 hours.

These variations in light intensity and duration of illumination affect photosynthesis directly. The periodicity of illumination also has indirect effects on
photosynthesis as a result of photoperiodic responses which determine the
form and life-cycle of the plant. Variations in the spectral composition at the
Earth's surface are perhaps not of great significance for present purposes.
The changes in the spectral composition of light which occur as it passes
through water are more important for photosynthesis, but discussion of this is
best left until we consider the absorption of light in the next chapter.

Carbon dioxide in aquatic environments

Carbon dioxide dissolved in water reacts to give carbonic acid and this dis-
sociates to give bicarbonate and carbonate ions, the various forms being in
equilibrium:

13) $H_2O + CO_2 \rightleftharpoons H_2CO_3 \rightleftharpoons H^+ + HCO_3^- \rightleftharpoons 2H^+ + CO_3^{--}$

At a given temperature and pressure the total amount of carbon dioxide and
its distribution among these forms depends on pH and the amount of "excess
base", that is of cations over and above that required to neutralize anions
other than carbonate and bicarbonate. The total amount of inorganic carbon
is consequently greatest in alkaline waters with high concentrations of excess

base, such as sea-water and "hard" fresh waters, and least in acid fresh waters. The variation with pH in amounts of the various forms in sea-water is illustrated in Fig. 3.

It is important to know which of these forms of carbon is utilized by aquatic plants. Whether free carbon dioxide or bicarbonate ion is directly concerned in the carboxylation reaction of the carbon fixation cycle (see p. 64) is uncertain, but for most purposes this question is unimportant since all photosynthetic cells so far tested contain an enzyme, carbonic anhydrase, which catalyses the conversion of one to the other. We are more concerned with the form in which carbon dioxide enters the cells. This may be determined by observing the effects on photosynthesis of varying concentrations of the different

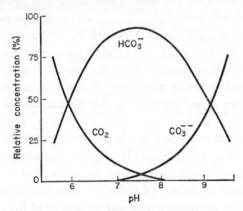

Figure 3 The relative concentrations of carbon dioxide, bicarbonate ion and carbonate ion at different pH values in sea-water of 34‰ salinity at 18°C. (Redrawn from E. Paasche (1964), *Physiol. Plant. Suppl.* III.)

forms, which may be achieved by altering the pH provided that due correction is made for any direct effect of pH on rate of photosynthesis. It is well established that all aquatic plants, both algae and higher plants, use free carbon dioxide, together with undissociated carbonic acid. Some, but not all, may also take up bicarbonate ion. The water moss, *Fontinalis antipyretica*, is unable to use bicarbonate as a direct source of carbon dioxide but aquatic flowering plants such as *Elodea canadensis* and *Potamogeton lucens* are able to. A curious feature is that the uptake, which takes place in the light, is accompanied by the excretion of OH^- ions on the dorsal side of the leaf with a resultant increase in pH there and precipitation of calcium carbonate. The mechanism of this process is not fully understood. Among algae *Chlorella pyrenoidosa* seems only to take up free carbon dioxide, but a related alga *Scenedesmus quadricauda* is able to use bicarbonate directly after a preliminary adaptation period in the light. Since many kinds of algae are able to photo-synthesize at appreciable rates at a pH as high as 11, at which the concentra-tion of free carbon dioxide is extremely low, it is likely that this ability to utilize bicarbonate ion is widespread. Some algae, like the higher plants men-tioned above, precipitate calcium carbonate and, possibly, this too is con-nected with bicarbonate utilization. A marine flagellate, *Coccolithus huxleyi* which produces elaborate scales, called *coccoliths*, constructed of calcium car-bonate, during photosynthesis has been shown to utilize both free carbon dioxide and bicarbonate ion and the indications are that a strain of this organism unable to produce coccoliths is also unable to use bicarbonate. Experiments with the inhibitor CMU (see p. 52) suggest that coccolith form-ation may be dependent on photophosphorylation and not directly linked to photosynthetic carbon assimilation so it is certainly a more complicated process than mere precipitation at high pH. Coccolith formation by *Cocco-lithus huxleyi* and its allies is geochemically perhaps the most important cal-cium carbonate forming process and more than half of the vast geological deposits of this substance in the form of chalk comes from this source. There is no evidence of any plant being able to make direct use of carbonate ion.

Diffusion in aqueous solution is much slower than it is in air. The diffusion coefficient of carbon dioxide in water is only one ten-thousandth of that in air and that of bicarbonate ion is about an eighth of that of carbon dioxide. Consequently it is often the rate of diffusion up to the site of reaction in the plant rather than the concentration of the substance in the bulk of the water which limits the rate of photosynthesis. In experimental work diffusion gradients in the medium in which the plant is immersed can be largely eliminated by stirring, but the path from the surface of the cell wall up to the chloroplasts is inaccessible to stirring and may be sufficiently long for diffusion to become limiting at high rates of photosynthesis. With *Chlorella*, which has spherical cells about 5 μ in diameter and a diffusion path from outside to the chloroplast of less than 1 μ, curves showing the relationship of photosynthesis to carbon dioxide concentration may be obtained which approximate to the form which would be expected were the reaction a simple enzymatic one. That is to say, at low concentrations the rate of photosynthesis is found to be proportional to concentration, whereas at high concentrations saturation is approached and the rate becomes independent of concentration (Fig. 4). In more elaborate aquatic plants, however, the diffusion path inside the cells is much longer—as much as 40 μ in *Myriophyllum spicatum*, for example. With such plants curves like that for *Chlorella* are obtained at low light intensities, when carbon dioxide is not limiting, but at higher intensities no carbon dioxide saturation is achieved (Fig. 5). This is because carbon dioxide is then consumed so rapidly that a saturating concentration cannot be maintained at the site of reaction and the rate of photosynthesis becomes dependent on the

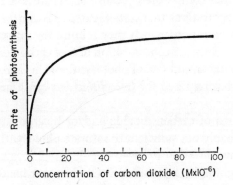

Figure 4 The effect of carbon dioxide concentration on photosynthesis by *Chlorella* at high light intensity. (Based on data of E. Warburg (1919), *Biochem. Z.*, **100**, 230.)

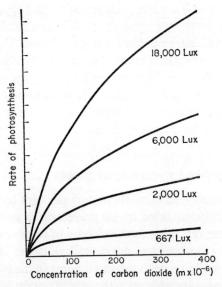

Figure 5 The relationship of rate of photosynthesis by the aquatic moss, *Fontinalis antipyretica*, to carbon dioxide concentration at different light intensities. (Redrawn after R. Harder (1921), *Jb. Wiss. Bot.*, **60**, 531.)

rate of supply by diffusion. This rate is proportional to the concentration gradient, $(C - C_s)/L$, where C is the concentration supplied, C_s that where the reaction takes place and L the length of the diffusion path.

It will be realized from this that the relationship of rate of photosynthesis to carbon dioxide concentration is very much dependent on the level of other factors. Certainly there is no such thing as a fixed optimum concentration. The situation is further complicated because plants are able to adapt to altered carbon dioxide concentrations to a certain extent. Nevertheless species seem to differ in the concentrations which they require for saturation under comparable conditions. Thus the concentration of free carbon dioxide (0·01 mM) required to give the maximum rate of photosynthesis in *Chlorella* is only about one-hundredth of that required for *Coccolithus huxleyi*, a marine flagellate of comparable size.

The concentrations of carbon dioxide and/or bicarbonate in fresh waters and the sea are probably not sufficient to saturate photosynthesis at high light intensities. The concentration in sea-water limits photosynthesis by *Coccolithus huxleyi* in bright light, for example. On the other hand the total reserve of inorganic carbon in natural waters is often high and the concentrations of carbon dioxide itself and bicarbonate will be maintained by equilibration between the various forms. Only when photosynthesis is exceptionally rapid is exhaustion approached. This has been recorded as happening in a Danish lake in which algal growth was especially prolific as a result of enrichment of the water with mineral nutrients from sewage effluent. The pH of the water rose as high as 10·2 as a result of removal of carbon dioxide by the plants and the total photosynthesis in the lake became limited by the rate at which it was taken up from the air at the lake surface. It was estimated that between 60 and 100 g of carbon dioxide were absorbed per square metre of surface per month at the height of the summer season.

Carbon dioxide uptake by terrestrial plants

The atmosphere normally contains about 0·03 per cent by volume of carbon dioxide. Gaseous diffusion aided by air movement is sufficient to maintain the concentration near this value in the vicinity of plants except when photosynthesis is especially vigorous and the air still. Entry of carbon dioxide into flowering plants is largely through the small pores in the epidermis known as stomata, and when these are wide open the rate of uptake is almost as much as it would be if gaseous exchange could take place over the entire outer surface of the leaf. Brown and Escombe (1900) working at Kew showed by investigations on the diffusion of gases through small pores that although stomata when fully open occupy less than 1 per cent of the leaf surface, gaseous exchange through them may nevertheless be almost as much as if it were taking place over the whole leaf surface. This effect found theoretical explanation in the work of the physicist Stefan (1882) and the mathematician Jeffreys (1918). Putting it in its simplest terms, diffusion through a small hole in an impervious barrier is maintained by molecules from a much larger effective space than is diffusion through an equivalent part of a large plane area through which diffusion is occurring. The rate of diffusion of carbon dioxide into the leaf is controlled by the stomatal aperture but the effect depends on light intensity (Fig. 6). At low intensities there is scarcely any restriction on assimilation of carbon dioxide until the stomata are nearly closed, whereas at high intensities, when carbon dioxide is limiting, assimilation continues to increase

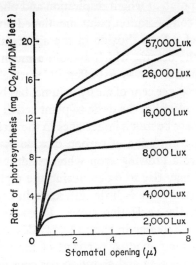

Figure 6 The relationship of rate of photosynthesis by leaves to different light intensities and stomatal opening. (Redrawn after Ståfelt (1960), *Encyclopedia of Plant Physiology*, **5** (2), 81; Springer-Verlag.)

until the stomata are fully open. Complete closure of the stomata virtually stops entry of carbon dioxide into the leaf. Inside the leaf carbon dioxide diffuses in the intercellular air-space system (Plate V) and dissolves in the water imbibed in the cell walls. To a certain extent movement of carbon dioxide in the intercellular system will be assisted by mass movements of air caused by temperature and pressure changes and by mechanical deformation of the plant in wind. Carbon dioxide produced by respiration in the roots and colourless parts of the plant may find its way through the intercellular system into the leaves and so provide a supply supplementary to that obtained from the air. From the cell wall carbon dioxide diffuses to the chloroplasts, a distance of several microns (Plate V), which because of the slowness of diffusion in the aqueous phase offers more resistance to the movement of the carbon dioxide than the entire passage through the gaseous phase. Protoplasmic streaming may, however, assist the movement of carbon dioxide up to the chloroplasts.

Even if a plant is exposed to bright light in the absence of a continuing supply of carbon dioxide, the concentration of this gas in the intercellular spaces is not reduced below 0·005 or 0·0005 per cent (see page 90). Placed in air containing less than this concentration, the carbon dioxide in the inter-cellular spaces increases as a result of respiration until the same residual concentration is attained. In some ways this can be looked on as a carbon dioxide compensation point, at which respiration and photosynthesis balance, analogous to the light compensation point mentioned on page 5 (Fig. 2).

The concentration of carbon dioxide in the air remains remarkably constant largely because the oceans, which contain a much larger total amount than does the atmosphere, act as a buffer. Nevertheless, a small annual variation, of about 0·0004 per cent of air by volume, has been detected in temperate latitudes and is related to the cycle of plant activity. Superimposed on these variations there may be much larger ones in the vicinity of vegetation in calm weather. The concentration of carbon dioxide may drop to 0·028 per cent immediately above a growing crop when conditions are favourable to photosynthesis, and it may rise to 0·040 during the night. In most weather mixing is so vigorous that there is little difference in concentration between the free air and within the vegetation cover. Release of carbon dioxide by decomposition of organic matter in the soil provides an appreciable supply for photosynthesis, forming about 6 per cent of the net carbon uptake for rapidly growing grass in spring and about 20 per cent for other crops in the summer according to experiments carried out at Rothamsted Experimental Station in southern England.

The rate of photosynthesis of terrestrial plants may be increased if higher than normal concentrations of carbon dioxide are supplied. This is, of course, only effective if light intensity or other factors are not limiting. Thus yields of spinach beet grown in a glasshouse can be increased five-fold by raising the carbon dioxide concentration to 0·60 per cent, but similar concentrations have smaller or no effects in winter. Concentrations of this order may cause damage. It has been recorded that leaves of sugar beet and tomato plants were killed after the plants had been cultivated for one or two weeks in an atmosphere containing 0·20–0·30 per cent of carbon dioxide, but such effects depend very much on the level of other factors such as light intensity and nutrition. Adaptation by growth in progressively higher concentrations enables a plant to withstand the damaging effects more successfully. Carbon dioxide enters directly into a variety of enzymic reactions and affects others indirectly by causing alterations in pH, so that it is likely that the damage results from a general derangement of metabolism.

Photosynthesis and the water relations of the plant

Photosynthetic tissues contain so much water—normally 80 to 90 per cent of their fresh weight—that it is unlikely that water concentration can ever be a directly limiting factor in photosynthesis. Leaves that lose sufficient water to become wilted do actually show a marked decline in photosynthetic activity, but this must be ascribed to structural and metabolic derangement rather than direct limitation. Chloroplasts isolated from wilted leaves show greatly reduced photochemical activity as compared with those from corresponding turgid leaves.

Nevertheless, photosynthesis and the water economy of a terrestrial plant are so intimately interrelated that water supply is normally one of the most important factors determining the amount of photosynthesis carried out.

Open stomata offer as little obstruction to the outward diffusion of water vapour as they do to the inward diffusion of carbon dioxide. The large expanse of damp cell surface presented to the air inside a leaf makes for efficient absorption of the carbon dioxide required for photosynthesis but it has the grave disadvantage that, except in a saturated atmosphere, it is equally efficient in evaporating water. The extent to which this water loss is controlled by stomatal closure has long been a matter of argument, but since about 1948 evidence has been accumulating that stomata do regulate evaporation from the leaf rather effectively in moving air. Closure of the stomata stops the entry of carbon dioxide as effectively as it prevents the escape of water vapour. Except when plants are grown in a saturated atmosphere the amount of water actually consumed in the synthesis of organic matter is a small fraction, one per cent or less, of that lost by evaporation. This loss and the movement of water through the plant which it brings about serve no essential physiological ends which could not be accomplished in other ways and it must be regarded as the inevitable price which a plant has to pay for carrying out photosynthesis in a terrestrial habitat. Since the supply of water available to a plant is usually limited—Penman has shown that in SE England crops would benefit from irrigation in eight years out of ten—successful growth depends on a balance being struck between photosynthesis and conservation of water.

The stomata are the main means of striking this balance. These structures respond to changes in several factors. They open if the concentration of carbon dioxide in their vicinity falls below 0·03 per cent and also in the light,

B

provided that other factors are not operating to make them close. Stomata close in the dark and under conditions of water-shortage. They thus respond in ways which not only will promote photosynthesis but will reduce water loss. Generally speaking, the effects of stomatal opening and closing on carbon dioxide intake and water vapour loss are parallel but there is one important exception to this. In still air stomata have to be almost closed before they exert any appreciable control on water loss, the chief resistance to evaporation then being offered by the layer of water-saturated air which covers the leaves and constitutes a "diffusion barrier". Under these circumstances carbon dioxide absorption, which is not, of course, affected by the concentration of water molecules in the air, may be increased, if the light is sufficiently bright, by stomatal opening without any concomitant increase in water loss. In wind when the diffusion barrier is blown away increased carbon dioxide absorption can only be achieved by stomatal opening at the cost of a corresponding water loss.

The interaction of the various responses of stomata together with less well-marked effects of temperature and an inherent daily cycle of opening and closing which occurs even under constant conditions, results in a pattern of behaviour which is complicated and difficult to predict. With ample water supply the stomata of some plants, the onion and potato for example, remain open all the time, but if water is short, they usually close at midday and remain closed for much of the night. The stomata of cereals remain closed at night and even under favourable conditions open for only an hour or two in the day. In succulent plants on the other hand they open at night and close in the day.

These apparently capricious patterns of behaviour do seem, on closer inspection, to be such as to prevent water loss as far as possible and at the same time allow some photosynthesis. Midday closure restricts water loss when the evaporating power of the air is reaching its maximum, and closure at night saves a small water loss without interfering with photosynthesis. The succulents, which are characteristic of habitats in which water is usually in extremely short supply, have evolved a biochemical mechanism which enables them to take carbon dioxide in and store it as organic acid during the night when water loss by evaporation is at a minimum. During the day, behind tightly shut stomata, this stored carbon dioxide can then be photosynthesized into sugars without loss of water to the air.

Perhaps stomatal control may be even more precise than these generalizations would suggest. Studies on stomatal opening at various levels within agricultural crops suggest that the amount of light reaching the lower leaves is the crucial factor determining their opening, so that it appears that a leaf not receiving sufficient light for photosynthesis does not open its stomata and thereby limits unnecessary water loss.

Further Reading

Bainbridge, R., Evans, G. C., and Rackham, O. (Editors) (1966) *Light as an Ecological Factor*. Blackwell, Oxford.

Becker, C. F., and Boyd, J. S. (1957) Solar radiation availability on surfaces in the United States as affected by season, orientation, latitude, altitude and cloudiness. *J. Solar Energy Sci. Engng*, **1**, 13–21.

Lemon, E. (1967) Aerodynamic studies of CO_2 exchange between the atmosphere and the plant. In *Harvesting the Sun* edited by A. San Pietro, F. A. Greer, and T. J. Army, pp. 263–90. Academic Press, London and New York.

Meidner, H., and Mansfield, F. A. (1968) *The Physiology of Stomata*. McGraw-Hill, London.

Montieth, J. L., Szeicz, G., and Yabuki, K. (1964) Crop photosynthesis and the flux of carbon dioxide below the canopy. *J. appl. Ecol.*, **1**, 321–37.

Nichiporovich, A. A. (Editor) (1967) *Photosynthesis of Productive Systems*. Israel Program for Scientific Translations, Jerusalem.

Paasche, E. (1964) A tracer study of the inorganic carbon uptake during coccolith formation and photosynthesis in the coccolithophorid *Coccolithus huxleyi*. *Physiol. Plant. Suppl.* III, 5–82.

Rabinowitch, E. I. (1951) *Photosynthesis and Related Processes*, vol. II part 1. Interscience, New York.

Stålfelt, M. G. (1960) Allgemeinere Physiologie und Ökologie der Photosynthese. B.2. Das Kohlendioxyd. In *Encyclopedia of Plant Physiology* edited by W. Ruhland, vol. V (2), pp. 81–99. Springer-Verlag, Berlin.

Steemann Nielsen, E. (1960) Uptake of CO_2 by the plant. In *Encyclopedia of Plant Physiology* edited by W. Ruhland, vol. V (1), pp. 70–84. Springer-Verlag, Berlin.

3   *The absorption of light*

Properties of the plant pigments

IN order to have photochemical effects light must be absorbed rather than reflected or transmitted and, as we saw in the first chapter, the substances responsible for absorption in photosynthesis are chlorophyll *a* and the pigments associated with it in the chloroplasts. When extracted from the plant into solution these pigments absorb light according to the Lambert–Beer law, which may be expressed in the form

14) $$I_t = I_0 . 10^{-\alpha c d}$$

where I_t is the intensity of light after having passed through a thickness of solution, *d*, while I_0 is the incident intensity, α is the extinction coefficient of the absorbing pigment and *c* its concentration. This is to say that each successive equal thickness of solution absorbs an equal fraction of the light, regardless of its intensity (this is expressed in graphical form in Fig. 11), and the absorption is proportional to the number of absorbing molecules. The extinction coefficient, α, and therefore absorption also, varies with the wavelength of the light. The chlorophylls have maximum absorption in the blue-violet and red and the carotenoids in the blue-violet. The biliproteins (water-soluble proteinaceous accessory pigments which occur in the red seaweeds, blue-green algae and the flagellate group, the Cryptophyceae) are of two main sorts, the phycoerythrins, red pigments with maximum absorption in the blue-green wavelengths, and the phycocyanins, blue pigments with maximum absorption in the yellow-red.

The *absorption spectra* of these various pigments, the forms of which are characteristic of the particular substance and can be used as a means of identification, are shown in Figs. 7 and 8. The positions of the absorption peaks vary slightly according to the solvent used. The height of the curve at any particular point is a measure of the probability that a quantum corresponding to that particular wavelength will be absorbed. It will be seen that absorption by the photosynthetic pigments of green plants and algae falls

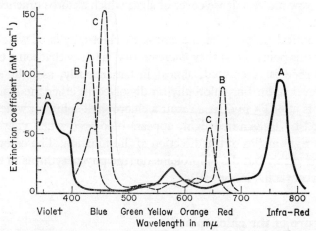

Figure 7 Absorption spectra of (A) bacteriochlorophyll, (B) chlorophyll *a*, and (C) chlorophyll *b*; all in ether.

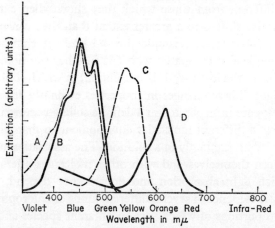

Figure 8 Absorption spectra of accessory pigments; (A) *β*-carotene in hexane, (B) fucoxanthin in hexane, (C) B-phycoerythrin from *Smithora naiadum*, in aqueous phosphate buffer solution, and (D) C-phycocyanin from the same alga in saline solution. (C and D redrawn from C. ó h Eocha (1960) in *Comparative Biochemistry of Photoreactive Systems*, Academic Press, p. 185.)

off at wavelengths shorter than 400 mμ and longer than 700 mμ. Thus only radiation of wavelengths between these limits is available for photosynthesis by these plants. This happens to coincide with the range of wavelengths visible to the human eye. Bacteriochlorophyll has an absorption peak extending down to 800 mμ, corresponding with the ability of the purple bacteria to utilize the near infra-red for photosynthesis. This enables these bacteria to grow under a dense cover of algae which absorbs practically all the visible light.

Another optical property of the extracted chlorophylls of which we may take note at this point is that they *fluoresce*, that is to say that some of the light energy absorbed is re-emitted, almost instantaneously, as light of a characteristic wavelength. For chlorophyll *a* dissolved in ether the main fluorescence band is at 664·5 mμ. As a result a chlorophyll solution, which is green when viewed by transmitted light, appears blood-red when viewed from a direction at right angles to the direction of illumination. The carotenoids do not fluoresce visibly but the biliproteins do (the phycoerythrins in the orange and the phycocyanins in the red).

The structure of the chloroplast

A chloroplast is far more than a sac containing a solution of photosynthetic pigments. Moreover, the properties of these pigments in the living cell are somewhat different from those which they show when extracted. The absorption peaks are shifted to a greater extent than they can be in solution by changing the solvent; for example the red peak of chlorophyll *a* is at 678 mμ in a living leaf as compared with 662 in ether solution. Chlorophyll is also more stable in the leaf than it is in solution, in which it bleaches rather quickly in sunlight. These changes in properties evidently arise because the pigment molecules are incorporated in a definite solid structure in the chloroplast, which gives a different molecular environment to that which is found when they are moving randomly in a solution. The association of chlorophyll molecules between themselves and with other molecules alters the electron orbitals on which light absorption and fluorescence depend. There is, in fact, spectroscopic evidence of several different kinds of chlorophyll *a* complex in plants, the proportions of which vary in different species and in the same species according to the conditions under which it is grown (see p. 49).

In higher plants chloroplasts are discoid or ellipsoidal bodies 5 to 10 μ in diameter. In the algae they may be larger and more complicated in shape. Even with the ordinary light microscope it can be seen that chloroplasts have internal structure—the pigment in higher plant chloroplasts appears concentrated in granules, the *grana*, embedded in a colourless *stroma*. Algal chloroplasts do not generally have grana (Plate XI). Study of the optical properties of chloroplasts in the 1930s showed that the various kinds of molecules must be arranged in definitely orientated platelets. It is reassuring that when the electron microscope was brought to bear on chloroplasts this picture was confirmed. Electron micrographs of moderate magnification (see Plate VIII) show that the chloroplast has a bounding membrane and that it contains electron-opaque grana, showing a layered structure, dispersed in a less opaque stroma, containing less closely spaced membranes. At higher magnifications (Plate X) the chloroplast membranes are seen to be in the form of flattened closed sacs, called *thylakoids,* which are closely stacked like piles of pancakes in the grana and which occasionally extend through the stroma, linking granum with granum. In sections examined at very high magnifications the membranes appear to be built of nearly spherical sub-units about 60 Å in diameter. When examined in face-view, however, the thylakoids seem to consist of arrays of particles, about 150 Å in diameter which have been called *quantasomes.* Possibly four of the 60 Å sub-units make a quantasome. It is tempting to believe that these structural units revealed by the electron microscope correspond with the photosynthetic units predicted from kinetic studies, but they may be nothing more than artefacts produced by the fixation procedure. In certain red seaweeds and blue-green algae regularly spaced granules, about 350 Å in diameter, have been found attached to the lamellae (Plate XI). These appear to contain the biliproteins. No characteristic supra-molecular structures have yet been demonstrated in the stroma (see Plates IX and X).

The photosynthetic organelles of bacteria, called chromatophores, are also essentially lamellar in structure. Sometimes they take the form of stacks of thylakoid-like structures but in *Rhodopseudomonas spheroides* these are swollen into nearly spherical vesicles.

The electron microscope gives little clue to the chemical nature of the structures which it reveals, but experimental evidence shows that the pigments are associated with the thylakoids and from a knowledge of chloroplast composition guesses can be made of the way in which the structure is built up. Chlorophylls and carotenoids make up from 5 to 10 per cent of the dry weight of the chloroplast, protein 41 to 55 per cent, and lipides 18 to 37 per cent. From Fig. 1 it will be seen that the chlorophyll molecule consists of a flat "head" built of subsidiary ring structures, with a magnesium atom in the centre, and a long phytol "tail". It is found that the amount of chlorophyll in a chloroplast is such that it could just be accommodated in the thylakoid membranes with the "heads" of the molecules lying flat on the membrane surfaces and the "tail" extending at right angles to this. However, from the optical properties of the chloroplast it can be deduced that the arrangement is not as regular as this, not more than 5 per cent of the chlorophyll *a* molecules being orientated in the same way. The arrangement of the various sorts of molecule in the sub-units is so much a matter of speculation at the moment that it does not seem worth while discussing the various structures that have been proposed. Apart from those substances already mentioned, the lipides, some of which, the galactosyl dilinolenins for example, are uniquely associated with the lamellar system, evidently play an important part in determining the molecular structure of the chloroplast. In the lipide layers the pigment molecules form a continuous array which is efficient in trapping light and the water-imbibed protein layers sandwiched between them provide a phase in which the enzyme reactions occur and the raw materials and products are transported.

Transfer of absorbed energy between pigments

An important feature of light utilization in photosynthesis depends on the way in which the pigment molecules are arranged in the chloroplast. If diatoms are illuminated with blue-green light (470 mμ), three-quarters of which is probably absorbed by the carotenoids (mainly fucoxanthin), the yield of chlorophyll fluorescence is much the same as when red light, absorbed exclusively by chlorophyll, is used. This suggests that energy

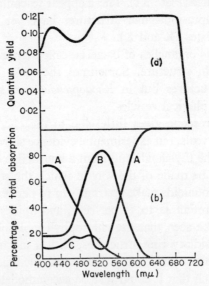

Figure 9   (*a*) Quantum yield of photosynthesis of the diatom *Navicula minima* as a function of wavelength. (*b*) The estimated distribution of light absorption among pigments in living cells of *Navicula minima* as a function of wavelength; A, chlorophylls *a* and *c*, B, fucoxanthin, and C, other carotenoids. (Redrawn from T. Tanada (1951), *Amer. J. Bot.*, **38**, 276.)

absorbed by the fucoxanthin is transferred to the chlorophyll from which some of it is emitted as fluorescence. That energy absorbed by accessory pigments such as fucoxanthin can be used for photosynthesis has been shown particularly clearly for the diatom *Navicula minima*. The quantum yield of photosynthesis was measured at different wavelengths (Fig. 9a) and compared with the distribution of the absorbed light between chlorophylls, fucoxanthin and other carotenoids (Fig. 9b). Because of the considerable difference in the absorption peaks *in vivo* as compared with those in solutions of extracted pigments, the estimation of the distribution of absorbed light was not straightforward but was accomplished by measurement of the absorption spectra of cells from which fucoxanthin and chlorophyll *c* had been pre-ferentially extracted. The quantum yield was found to be nearly constant from 520 to 680 mμ, falling abruptly at longer wavelengths and showing a slight dip in the blue between 430 and 520 mμ. Comparison of this curve with that for the distribution of absorption among the pigments *in vivo* shows that at 550 mμ, where some 80 per cent of the light absorption is by fucoxanthin, the efficiency of photosynthesis as measured by the quantum yield is the same as that at 650 mμ where 100 per cent of the light is absorbed by chlorophylls. Light absorbed by the carotenoids other than fucoxanthin does not appear to be available for photosynthesis.

In a wide variety of other algae the availability for photosynthesis of light energy absorbed by certain accessory pigments has been demonstrated by comparison of absorption curves and action spectra. An action spectrum for photosynthesis is obtained by measuring the photosynthetic yield for a given amount of light incident upon the plant over a range of wavelengths. The measurements must, of course, be made at low light intensity so that the amount of photosynthesis is proportional to the light intensity. The example given in Fig. 10 is for *Phormidium ectocarpi*, a blue-green alga. Besides chlorophyll *a* and various carotenoids this alga contains *C*-phycoerythrin, the pronounced peak in the absorption spectrum in Fig. 10 at 567 mμ being due to this latter pigment. It will be seen that except in the region 450 to 520 mμ, where the carotenoids are responsible for much of the absorption, the absorption and action spectra parallel one another closely. Since in the region of 567 mμ the other pigments absorb scarcely at all this implies that light absorbed by *C*-phycoerythrin is available for photosynthesis.

It thus appears that light energy absorbed by one kind of pigment may be transferred to another and that light energy absorbed by some, but not all, of the accessory pigments is as useful in photosynthesis as that absorbed by chlorophyll itself. Various well-attested physical mechanisms for such trans-fer are known but it is difficult to decide which operates in the chloroplast.

One possibility is that the transfer is effected by resonance, a relatively slow process in which radiation is not involved, occurring when molecules are closely juxtaposed in regular orientation. A condition of such a transfer is that the receiving molecule should have an absorption band overlapping with that of the other at the same or a longer wavelength. Another possibility

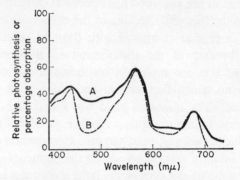

Figure 10 Absorption spectrum of the living cells (A) and photosynthetic action spectrum (B) of the blue-green alga *Phormidium ectocarpi*. The two curves have been adjusted to coincide at 675 mμ. The absorption peak at 567 mμ is attributable to *C*-phycoerythrin. (Redrawn from F. T. Haxo (1960) in *Comparative Biochemistry of Photoreactive Systems*, Academic Press, p. 348.)

is that the different pigments are united into a crystal-like structure in which they share common electron orbitals. Light energy absorbed in any part would raise the energy of the whole of such a complex. Either mechanism would equally well account for the transfer of energy from one chlorophyll molecule to another as well as for the transfer from accessory pigment to chlorophyll.

The photosynthetic unit

We have seen that there is evidence from kinetic studies for the existence of a photosynthetic unit (p. 6) and that a structure, the quantosome, which may correspond with this, is distinguishable in electron micrographs of chloroplasts (p. 27). The concept of a photosynthetic unit is now indispensable in any picture of the mechanism of photosynthesis and perhaps we might at this point look at other evidence for its existence even though this involves anticipating to a certain extent what is to follow in later chapters.

A dense suspension of *Chlorella* begins to evolve oxygen immediately on illumination with weak light although, as can be shown by calculation, each chlorophyll molecule would under these conditions have to wait an average of an hour to collect the quanta necessary for the reduction of a molecule of carbon dioxide. The transfer of energy between pigment molecules which has just been described provides an explanation for the absence of a lag in oxygen evolution since it allows quanta collected by any of 2500 or so chlorophylls to be available for the reduction. The transfer is sufficiently rapid for this; it has been estimated that as many as 10,000 transfers can take place during the short space of time, 10^{-2} sec, which is the lifetime of the excited state of the chlorophyll molecule (p. 43).

The active centre where the absorbed energy is collected and utilized is probably concerned with the transfer of an electron rather than the direct reduction of an entire carbon dioxide molecule. On the assumption that this requires one quantum it seems likely that the photosynthetic unit comprises 200–400 chlorophyll molecules instead of the 2500 calculated on the basis of the requirements for carbon dioxide reduction. This is in accord with other findings. Thus if isolated chloroplasts are broken into progressively smaller fragments they lose their characteristic photochemical activity when the fragments contain 200 or less chlorophyll molecules. Some of the substances believed to be components of the active centre and concerned in electron transfer, for example the chlorophyll modification P 700 (p. 49), cytochrome *f* (p. 47) and ferredoxin (p. 48), are present in the proportion of one molecule per 400 or so of chlorophyll. Similarly it is found that one molecule of the specific photosynthetic inhibitor DCMU (p. 52) is sufficient to inactivate 200 chlorophyll molecules.

The structure of the photosynthetic unit is certainly more complicated than these facts would suggest since it must also allow for two distinct photochemical reactions (p. 55), and as yet we can only speculate about the mechanism which regulates and co-ordinates them. The working out of the detailed architecture of this minute, unique, and highly efficient energy conversion unit is surely one of the most exciting tasks in the whole of science.

The protective rôle of carotenoids

It was remarked above that light absorbed by certain carotenoids is unavailable for photosynthesis. Yet such carotenoids are of universal occurrence in photosynthetic organisms and it seems reasonable to suppose that they have a function. Perhaps this function is a protective one. In the presence of light and oxygen, photoxidations are likely to occur which can destroy the chlorophyll and it seems that carotenoids can prevent this. A mutant of *Rhodopseudomonas spheroides*, blue-green in colour, has been found which lacks the coloured carotenoids of the wild type. Under anaerobic conditions it shows normal photosynthetic growth but if exposed to light in the presence of oxygen its bacteriochlorophyll is photooxidized and it is killed. Strains of the bacterium possessing carotenoid pigments are not sensitive in this way. It is presumed that the carotenoids interact with potentially destructive triplet states of chlorophyll and divert the excitation energy into harmless paths. It should be noted that carotenoids are not indispensable, because the blue-green mutant of *R. spheroides* does function normally so long as no oxygen is present.

Light absorption and rate of photosynthesis by unicellular plants

Microscopic plants such as the plankton algae of the oceans which, as we shall see, produce a total photosynthetic yield at least equal to that of terrestrial vegetation, are almost completely at the mercy of their environment. Some species, swimming by means of flagella, can migrate slowly towards or away from a light source, but on the whole these unicellular plants must accept the light conditions in their vicinity.

Light penetrates into water to a greater or lesser extent according to the water's content of suspended matter and coloured dissolved substances. If

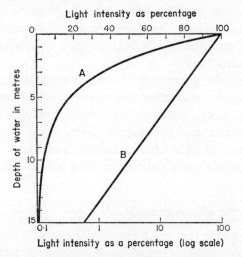

Figure 11 Penetration of light into moderately clear lake- or sea-water with uniformly distributed plankton. A, light intensity as a percentage of that at the surface; B, the same plotted on a logarithmic scale.

these materials are distributed uniformly then light penetration takes place according to the Lambert–Beer law so that the intensity decreases with increasing depth in an exponential manner as shown in Fig. 11. In clear ocean water sufficient light penetrates for the compensation point, that point where, over a twenty-four-hour period, photosynthesis just balances respiration, to be at a depth of 100 metres or so. In very turbid water, however, the compensation point may be at a depth of only a few centimetres.

By combining the curve for light penetration given in Fig. 11 with that showing the relation of rate of photosynthesis to light intensity (Fig. 2), we can predict how the rate of photosynthesis should vary with depth in a body of water containing a uniformly distributed algal population. Near the water surface in full sunlight the light intensity will be so high as to be inhibitory. Going down, the light intensity will decrease and the rate of photosynthesis will increase, reaching a maximum at a depth of 1 or 2 metres in moderately clear water (Fig. 12). At this maximum, photosynthesis is

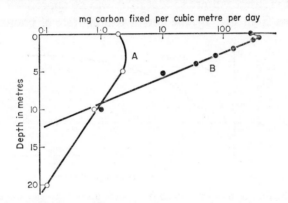

Figure 12 Photosynthesis in relation to depth in A, Torneträsk, an infertile lake in Swedish Lappland, and B, Erken, a fertile lake near Uppsala, both in late August 1956. Note that photosynthesis is plotted on a logarithmic scale. (Data of G. E. Fogg (1958), *Rapp. Proc.-Verb. Cons. internat. Explor. Mer.* **144**, 56.)

light saturated. Below this point, photosynthesis becomes light-limited and the curve of its rate against depth parallels that of light intensity against depth. On a dull day the surface illumination may already be below saturation so that the whole photosynthesis–depth curve resembles the intensity–depth curve. Actual measurements of photosynthesis in lakes and the sea agree closely with the theoretical curves provided that the algal cells are uniformly distributed; if they are not then the curve will be distorted. Since the cells themselves absorb light they will to a greater or lesser extent determine the penetration of light. In an infertile water the numbers of plankton will be low and absorption by non-living suspended matter, by dissolved substances and by the water itself will be the main determinants of light penetration. The great depth through which photosynthesis is possible in infertile water to some extent offsets the low rate of photosynthesis possible per unit volume of water. In fertile waters, however, the plankton algae will be self-shading so that the photosynthesis curve becomes drawn out into a prominent maximum near the surface, petering out rapidly below this (Fig. 12).

It is not completely true to say that the rate of photosynthesis of a particular kind of algal cell at a given depth is determined solely by light intensity. To a limited degree cells are capable of adaptation and of regulating the amount of light absorbed. With various species of algae it has been observed that cells grown in weak light have a higher pigment concentration than those grown in strong light. This results in different responses to different light intensities as has been demonstrated rather clearly for the green alga *Chlorella vulgaris*. The alga was grown under high (30 kilolux) and low (3 kilolux) intensity light in dilute culture so that there was no appreciable self-shading. As will be seen from Fig. 13, the curves relating rate of photosynthesis per

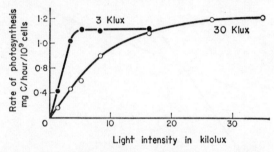

Figure 13   The rate of photosynthesis per unit (number of cells, in relation to light intensity, for *Chlorella vulgaris* grown at low (3 kilolux) or high (30 kilolux) light intensity. (Redrawn from E. Steemann Nielsen, V. K. Hansen and E. G. Jørgensen (1962), *Physiol. Plant.*, **15**, 508.)

cell to light intensity were distinctly different for these two cultures. The cells grown at low light intensity were more efficient at low intensity but became saturated at a lower level of illumination than the cells grown in intense light. The greater efficiency of the low light cells resulted from their higher concentration of chlorophyll since, if photosynthesis per unit amount of chlorophyll is plotted instead of photosynthesis per cell, there is no difference between the two cell types at limiting intensities. The low light and high light algal cells correspond roughly to the "shade" and "sun" leaves of higher plants which will be described below. Adaptation by *Chlorella* from one condition to the other is rather quick, requiring only one cell division cycle, which takes about seventeen hours under the particular conditions which were used. Nevertheless, if the waters of a lake or the sea are turbulent the plankton will be constantly moved from one depth to another and get no chance to adapt to any particular level of illumination. The photosynthesis-light intensity curves of plankton algae taken from near the surface and from deep down have, in fact, been found not to differ in the early winter when the water is well mixed. In summer, however, the water column becomes stabilized by thermal stratification and the algae do become adapted as Fig. 14 shows. Absorption of light may also be controlled to a certain extent by chloroplast orientation in green and brown algae and diatoms. This is

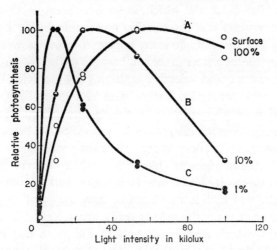

Figure 14 The rate of photosynthesis, in relation to light intensity, of phytoplankton taken from depths in the Sargasso Sea to which, A, 100%, B, 10%, and C, 1%, of the surface light penetrated. (Redrawn from J. H. Ryther and D. W. Menzel (1959), *Limnol. Oceanogr.*, **4**, 492.)

shown most clearly in the filamentous green alga *Mougeotia*, the flat plate-like chloroplast of which becomes orientated edge-on to intense light but face-on to moderate light.

As a light-absorbing system, a natural plankton population is, on the whole, inefficient. The cells at the surface receive so much light that they may be actually inhibited but the bulk of the population is light-limited. In artificial cultures of algae more efficient utilization of light may be achieved, but this is a matter for discussion later (p. 106).

Chromatic adaptation

Yellow and red predominates in sunlight as it is received at the earth's surface. Water is not completely colourless, even if pure, but absorbs yellow, red and violet light more than it does green and blue, with the result that sunlight is changed in spectral composition as it passes through water. It would therefore seem to be of advantage to an alga growing in deep water to possess photosynthetic pigments having an absorption maximum in the blue-green, as this would enable them to secure maximum absorption of the sparse light available. It is a fact that the predominating groups of marine plankton algae have a yellow-brown coloration in contrast to the bright green of plants which grow in unaltered sunlight. Thus the diatoms and coccolithophorids both have as their main accessory pigment fucoxanthin, which absorbs strongly in the blue-green and which, as we have seen, is effective in photosynthesis. The possession of fucoxanthin must undoubtedly be of advantage in clear waters but diatoms, at least, appear to grow as well with similar pigmentation in coastal waters in which, because of the presence of dissolved yellow organic matter and selective scattering by particles in suspension, yellow light usually penetrates further than blue-green light.

The adaptation of pigmentation to the quality of light prevailing in a particular habitat, *chromatic adaptation* as it is called, is, however, most evident in the seaweeds of a rocky shore. Here the green seaweeds, possessing as their predominating pigment chlorophyll, which absorbs strongly in the yellow and red, are mostly found near high-water mark. Lower down are the brown seaweeds, which, like the diatoms, have fucoxanthin as their main accessory pigment. Finally the red seaweeds, characterized by the possession of phycoerythrin, which absorbs strongly in the green and blue, are most abundant at and below low-water mark. The algae are thus positioned where the light quality, when they are submerged, is complementary to their pigmentation. Engelmann drew attention to this in 1883 and obtained evidence by means of his bacterial method (p. 2) that the light absorbed by the accessory pigments is useful in photosynthesis. His conclusion that chromatic adaptation is of biological advantage in enabling an alga to ensure maximum photosynthesis in the light conditions in which it grows was, however, questioned. It is only in recent years with the advent of more precise proof of the use in photosynthesis of energy absorbed by the accessory pigments that it has been generally accepted as correct. There are nevertheless frequent exceptions; algae often occur "out of place" on the shore. It will be seen from Fig. 7 that chlorophyll does have slight absorption in the green so that effective absorption of green light may be secured by a green alga by increase in concentration of this pigment only or by increase in thickness of the photosynthetic tissue, and it may thereby be enabled to photosynthesize efficiently in deep water. Conversely, the possession of phycoerythrin may not be of actual disadvantage to a red alga growing high up on the shore.

The possession of particular accessory pigments is characteristic of groups of algae and is evidently genetically determined. The relative proportions of the various pigments in a given species may, however, vary. This occurs

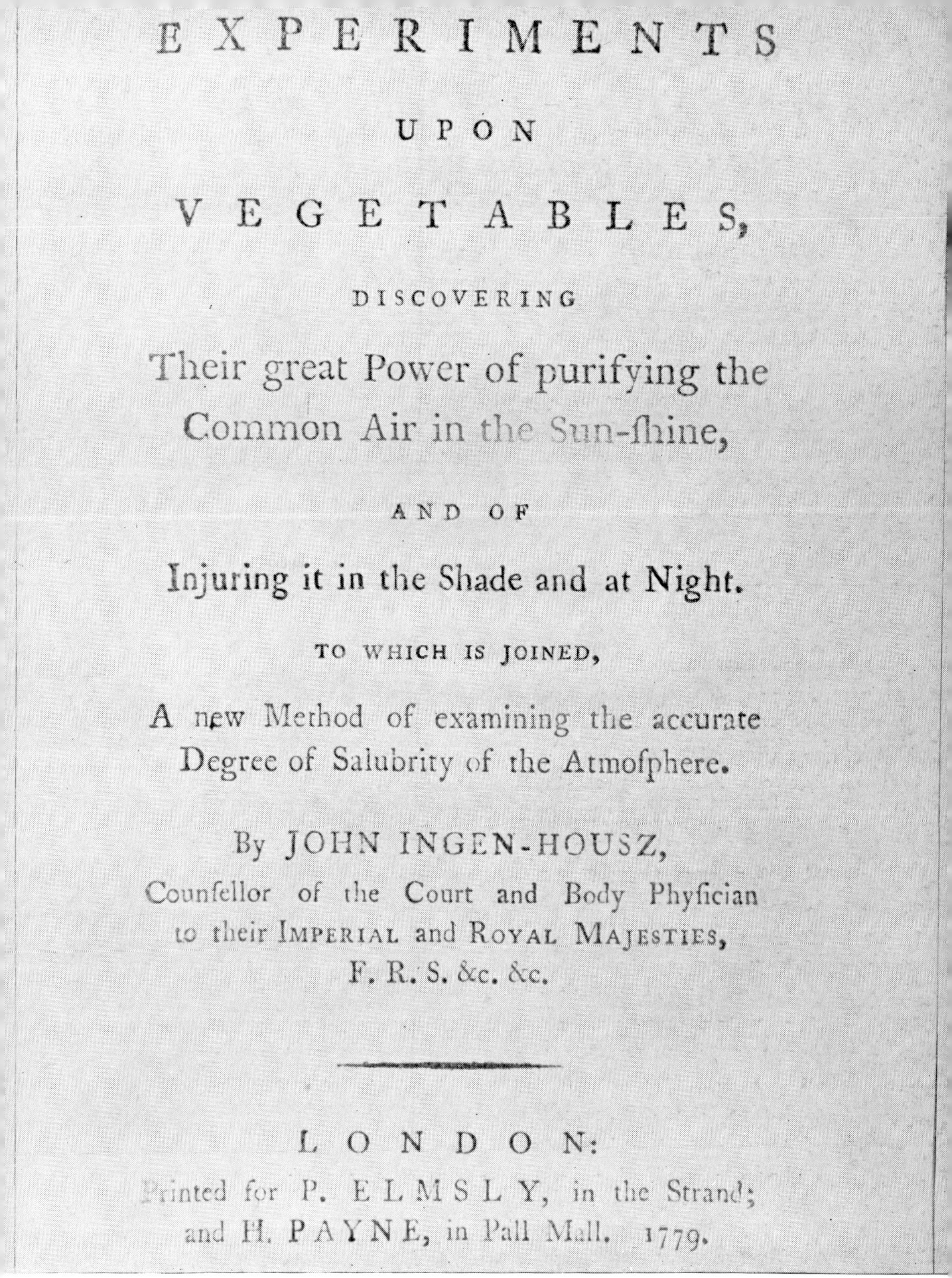

EXPERIMENTS

UPON

VEGETABLES,

DISCOVERING

Their great Power of purifying the
Common Air in the Sun-shine,

AND OF

Injuring it in the Shade and at Night.

TO WHICH IS JOINED,

A new Method of examining the accurate
Degree of Salubrity of the Atmosphere.

By JOHN INGEN-HOUSZ,

Counsellor of the Court and Body Physician
to their IMPERIAL and ROYAL MAJESTIES,
F. R. S. &c. &c.

———————

LONDON:

Printed for P. ELMSLY, in the Strand;
and H. PAYNE, in Pall Mall. 1779.

Plate I   The title-page of the first published book to be devoted specifically to photosynthesis. From
the copy presented by Ingen-Housz to the Royal Society.

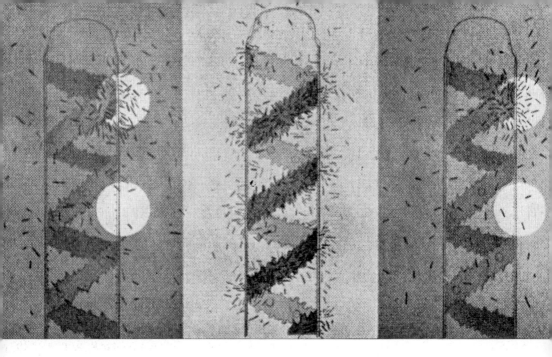

Plate II (*a*) Figures from an original paper by T. W. Engelmann (*Verh. Kon. Akad. Wetensch. Amsterdam*, Section 2, III (11), 1894) illustrating his bacterial method for studying photosynthesis. Left: the bacteria, which migrate towards higher oxygen concentrations, are seen congregated around an illuminated area of the chloroplast of the alga *Spirogyra* but not around a similarly illuminated region of colourless protoplasm. Centre: the distribution of the bacteria when the whole cell is illuminated. Right: the upper spot is of red light, the lower of green, showing that the bacteria are not being attracted by the green light which has been transmitted by the chloroplast.

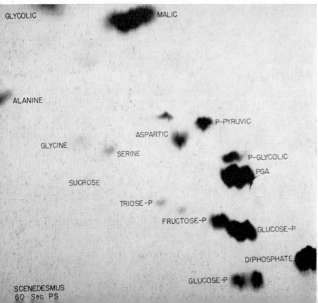

(*b*) Radioautograph of a two-dimensional chromatogram showing the labelled products in an extract from cells of the green alga *Scenedesmus* which had photosynthesized for 60 seconds in the presence of radiocarbon-labelled carbon dioxide. Reproduced by permission of the authors and publishers from fig. 3, J. A. Bassham and M. Calvin, *The Path of Carbon in Photosynthesis*, Prentice-Hall, 1957

Plate III (*a*) Sun leaves of *Pieris taiwanensis*. The upper leaves are nearly vertical and large gaps between them allow flecks of sunlight to reach the foliage lower down. (*b*) Shade leaves of *Sparmannia africana* (Tiliaceae). The large thin leaves are spread so as to intercept the maximum amount of light.

Plate IV The compass plant, *Silphium laciniatum*, growing in the Arboretum of the University of Wisconsin. The same plant (a) viewed from the west—leaves face-on. (b) viewed from the north—leaves edge-on.

notably in response to various factors such as the level of nutrients and light intensity but also to light quality. Thus chromatic adaptation may be phenotypic as well as genotypic. The blue-green algae show changes in pigmentation related to the wavelength of the light which illuminates them particularly well and in one species, *Tolypothrix tenuis*, the phenomenon has been investigated thoroughly. It is found that a quite short period of preillumination, as little as three minutes under suitable conditions, with light of a particular colour will quite markedly affect the proportions of the two biliproteins formed in a subsequent dark period. If the light is green, of wavelength around 500 mμ, then phycoerythrin which has an absorption peak at 576 mμ is mainly formed; on the other hand if the light is red, around 600–650 mμ, then phycocyanin with an absorption maximum of 630 mμ, predominates.

Light absorption and rates of photosynthesis by higher plants

Optically, as well as in other respects, a higher plant is a vastly more complicated system than the algae just considered. We may begin by considering light absorption by a single leaf. Its epidermal surface is usually shiny and in general about 10 per cent of the incident light is reflected. The light which penetrates enters tissues in which highly refractive and virtually colourless protoplasm and cell wall material abutting on air-spaces cause it to be reflected and refracted in a zig-zag path. The path length of light through a leaf is thus far greater than the actual thickness of the leaf and maximum opportunity is given for interception and absorption by the chloroplasts. Usually about 80 per cent of the incident light is absorbed, leaving about 10 per cent to be transmitted. The exact proportions reflected, absorbed and transmitted vary somewhat with wavelength, as is shown in Fig. 15. If this figure is compared with the absorption spectrum for chlorophyll given in Fig. 7 it will be seen that the absorption spectrum for the leaf in the visible has less pronounced peaks and troughs. This is because the concentration of the chlorophyll solution was chosen so as to accentuate the peaks, whereas in the leaf with an effectively high concentration of chlorophyll even green light is largely absorbed. The increase of reflexion and transmission and decrease of absorption of the leaf in the near infra-red is the cause of the white appearance of vegetation in photographs taken by infra-red radiation. Since infra-red cannot be used for photosynthesis by the green plant and has heating effects which are usually undesirable low absorption is of biological advantage. It is interesting that lichens and plants growing in frigid situations absorb infrared to a much greater extent than other plants.

As compared with an alga, which must absorb an almost fixed proportion of the light which falls upon it, whatever its intensity, a higher plant has a remarkable degree of control over the amount of light which it absorbs. This control is exerted in the first place by leaf orientation. The amount of light received by unit area of a plane surface is proportional to the sine of the angle of incidence so that a leaf edge-on to the sun's rays receives a minimum and one at right angles maximum amount. Full sunlight is inhibitory for photosynthesis so that when exposed to this it is an advantage if the leaf lies at an acute angle to the rays, since it will thereby effectively reduce the intensity to a level at which photosynthesis is more efficient and allow a

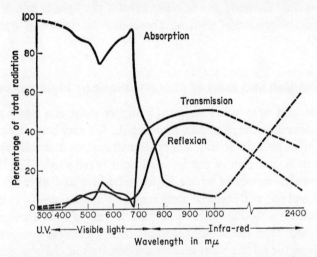

Figure 15 Reflexion, absorption and transmission of light by green leaves in relation to wavelength. (Redrawn from W. Tranquillini (1960), *Encyclopedia of Plant Physiology*, 5 (2), 304; Springer-Verlag.)

proportion of the light to pass on to be available for absorption by a leaf lower down (Plate III*a*). In the shade, on the other hand, it is of most advantage if the leaf lies at right angles to the general direction of illumination to intercept the maximum amount (Plate III*b*). In the Gramineae, which includes some of the photosynthetically most efficient plants such as sugar cane and maize, the growth habit results in the leaf surfaces being near vertical so that they are mostly lying at an acute angle to the incident rays when the sun is at its highest. The compass plant of the prairies, *Silphium laciniatum*, when grown in an exposed position has its leaves lying in a vertical north-south plane so that they avoid receiving full midday radiation but receive most in the morning and evening (Plate IV). Leaf orientation depends on growth habit and on geotropic and phototropic responses involving differential

growth or turgor changes in the petiole. Many species respond compara-
tively quickly to changes in intensity and direction of illumination. The
Sparmannia africana plant illustrated in Plate III*b*, for example, was capable
of twisting its leaves through 90° within twenty-four hours when the direction
of illumination was reversed.

The absorption of light by a leaf may be controlled to a certain extent by
movement of chloroplasts. In many plants the chloroplasts gather on the
illuminated front walls of the cells and orient themselves so as to present their
faces to the light in moderate intensities, whereas in bright light they line
the side walls with their axes parallel to the direction of illumination. The
transmittance of the leaves of the fern *Adiantum cuneatum* has been found to
increase by as much as a third as a result of such movement. The advantage
of this is probably avoidance of inhibition; no increases in rate of photosynthesis have been observed to follow such responses. When high or low
levels of illumination persist over the period of development of a leaf there
are largely irreversible responses in anatomy and pigment production which
result in sun and shade leaves being different. As compared with sun leaves,
shade leaves are larger, thinner and have larger chloroplasts and a higher
pigment content on a fresh weight basis.

Like the *Chlorella* cells grown in low intensity light, shade leaves photosynthesize more efficiently at low light intensities than do those grown in the
sun, but the rate of photosynthesis at light saturation is less. There are also
differences in intensity of respiration between sun and shade leaves so that
the compensation point (p. 32) for the latter may be as low as 0·3–1·0 per cent
of maximum daylight as compared with 1·3–7·5 per cent for the former.

The leaf arrangement, type of leaf and the total leaf area of a plant are
usually such as to promote maximum utilization of the light available. On a
deciduous tree such as a beech (*Fagus sylvatica*) for example, sun-leaves
orientated at acute angles to the general direction of illumination, are found
towards the crown. Apart from the small amount of light which they reflect
and transmit downwards, gaps between them allow a good deal of light to
pass directly to leaves below. Underneath are shade-leaves orientated at right
angles to the general direction of illumination and arranged in a leaf mosaic
which ensures maximum interception of the light. Because of these various
adaptations the response of an entire plant to increase in illumination is
different from that of a single leaf. For the latter the response is as shown in
Fig. 2 with saturation and then inhibition occurring as the intensity ap-
proaches that of full sunlight. With the entire plant saturation is not reached
even in full sunlight because those leaves which are shaded by others are
still in a limiting intensity of light.

It is possible to express these various factors in quantitative terms and use them in a mathematical model to predict rates of photosynthesis. A useful measure of the relative leaf area of a plant is the leaf area index (L), which is the ratio of its total leaf area to the ground area which it occupies. Leaf area index varies between 1 and 8 according to species and the habitat of the plant. For beech, which casts a particularly dense shade, L is some-times as high as 8. Montieth (1965) has determined, S, the fraction of light passing through unit leaf layer without interception, τ, leaf transmission, and the leaf area index, as defined above, and used them together with expressions describing the light-response curves of single leaves and the daily variation of radiation to calculate rates of photosynthesis. Rates calculated on this

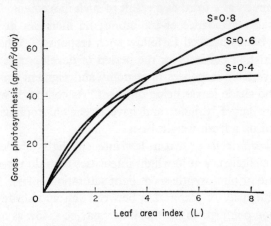

Figure 16 Calculated variation of gross photosynthesis in relation to leaf-area index (L) and interception factor (S) when the daily insolation is 600 cal cm^{-2}. (Redrawn from J. L. Monteith (1965), *Ann. Bot.*, **29,** 17.)

basis for crop-plants such as sugar beet, sugar cane, kale and subterranean clover are in good agreement with those actually observed. From this model it can be predicted that the same amount of photosynthesis should be achieved in a cloudy temperate climate with long days as in a more sunny equatorial climate with short days (see p. 101). It also gives precise expression to a point noted above, namely that at high light intensities greater rates of photo-synthesis are achieved by having a high leaf area index and a high value of S (which implies that the leaves stand more vertically). If the leaf area index is low more photosynthesis is achieved by having them arranged horizontally so that they intercept more light and S is low (Fig. 16).

Flowering plants have thus achieved a fairly good solution to the problem of utilizing a source of radiant energy which is highly variable and often too intense.

The effect of non-photosynthetic pigments in higher plants

Neither terrestrial higher plants, utilizing light which, compared with that available in aquatic habitats, is of fairly uniform spectral composition, nor the aquatic flowering plants and ferns which are descended from terrestrial ancestors, show anything comparable to the chromatic adaptation of the algae. Qualitatively the photosynthetic pigments of higher plants are always the same and their proportions vary relatively little except in autumn leaves when breakdown is occurring. Young and mature leaves of flowering plants do, however, frequently have pigments, occurring not in the chloroplasts but dissolved in the vacuolar sap, which give them a colour other than green. For example, the copper beech and the red and purple maples owe their colour to anthocyanins. Such pigments, of course, absorb light energy, but this is not available for photosynthesis and appears as heat. Their presence thus reduces photosynthetic efficiency at limiting intensities by absorbing light that might otherwise be absorbed by the chloroplasts. At saturating light intensities these non-chloroplastic pigments have no effect on the rate of photosynthesis.

Further Reading

Bainbridge, R., Evans, G. C., and Rackham, O. (Editors) (1966) *Light as an Ecological Factor*. Blackwell, Oxford.

Blinks, L. R. (1964) Accessory pigments and photosynthesis. In *Photophysiology* edited by A. C. Giese, vol. I, pp. 199–221. Academic Press, New York and London.

French, C. S., and Young, V. M. K. (1956) The absorption, action, and fluorescence spectra of photosynthetic pigments in living cells and in solutions. In *Radiation Biology* edited by A. Hollaender, vol. III, pp. 343–91. McGraw-Hill, New York and London.

Goodwin, T. W. (editor) (1966) *Biochemistry of Chloroplasts*, vol. I. Academic Press, London and New York.

Loomis, R. S., Williams, W. A., and Duncan, W. G. (1967) Community architecture and the productivity of terrestrial plant communities. In *Harvesting the Sun* edited by A. San Pietro, F. A. Greer and T. J. Army, pp. 291–308. Academic Press, London and New York.

Monteith, J. L. (1965) Light distribution and photosynthesis in field crops. *Ann. Bot.*, N.S. **29**, 17–37.

Nichiporovich, A. A. (Editor) (1967) *Photosynthesis of Productive Systems*. Israel Program for Scientific Translations, Jerusalem.

Rabinowitch, E. I. (1951) *Photosynthesis*, vol. II (1). Interscience, New York.

Talling, J. F. (1961) Photosynthesis under natural conditions. *Ann. Rev. Plant Physiol.*, **12**, 133–54.

4   *The conversion of light energy*

W E must now consider the events that lie between the absorption of light by the chloroplasts and the appearance of potential chemical energy, which we shall assume to be in the form of the two substances adenosine triphosphate, ATP, and reduced nicotinamide adenine dinucleotide phosphate $NADPH_2$. This phase of photosynthesis, the *photochemistry*, is the subject of intensive investigation at the present time and there is a wealth of apparently relevant observations. Nevertheless there remain large areas of ignorance and theories explaining the sequence of events must be regarded with caution.

For the understanding of this chapter it is necessary to appreciate the significance of electron transfers. Reduction is best defined as the acquisition of one or more electrons by a molecule, and this is usually equivalent to addition of hydrogen because it enables the molecule to capture hydrogen ions, always available in aqueous systems as a result of the dissociation of water. The reduction of fumaric acid to succinic acid, for example, may be represented as taking place in this way,

15) $$(CH \cdot COOH)_2 + 2H^+ + 2e \rightarrow (CH_2 \cdot COOH)_2$$

Interactions between light quanta and chlorophyll

Radiation physicists have provided much information about the events following the absorption of light by chlorophyll and similar substances *in vitro*. The light quanta, or photons, are only absorbed as whole units but absorption of a quantum may have one or more of several effects. One possibility is that it merely adds to the already existing thermal energy, which depends on translatory or vibratory movement of the whole molecule and which is not effective in photosynthesis. On the other hand, the potential energy of the absorbing molecule may be increased by displacement of an electron to an outer orbital.* The electrons in an atom tend to occupy the lowest energy levels; this is described as the *ground state*, as opposed to the *excited state* which is produced when an electron is displaced to an outer orbital. Four possible excited states of the chlorophyll molecule are represented in Fig. 17. These states are detected principally by spectroscopic study within extremely brief intervals following illumination. A red photon, absorbed in the band around 680 mμ, has a low energy content and produces the first excited *singlet* state, singlet being the state where the electron spins

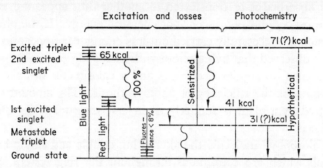

Figure 17 Diagram showing the energy levels in the chlorophyll molecule which may be of significance in photosynthesis. Energy-poor red light absorbed around 680 mμ produces the first excited singlet state. Energy-rich blue light absorbed around 430 mμ produces the second excited singlet state. The wavy lines denote unspecific heat losses. Conversion to the metastable triplet state competes with fluorescence, heat losses and photochemical conversions. (Redrawn from H. Gaffron (1960), in *Plant Physiology*, 1B, 50; Academic Press.)

are all paired, the spin of the excited electron continuing to be opposite to and neutralizing that of its ground state partner. A blue photon, absorbed in the band around 430 mμ, has higher energy content and produces a second excited singlet state. This decays within 10^{-11} sec to the first singlet state with liberation of part of its energy as heat. The energy of the first singlet state may be used for photochemical conversion, but there is little opportunity for this since its lifetime is only of the order of 10^{-9} sec. Another possibility is that it loses its energy by fluorescence and reverts to the ground state, or it may change to the metastable triplet state with a further small loss of energy as heat. This third state is termed *metastable* because it has a relatively long life of up to 10^{-2} sec and *triplet* because two electron spins are unpaired, the spin of the excited electron now being in the

* For a simple account of electron orbitals see *Energy, Life and Animal Organization*, by J. A. Riegel, English Universities Press, 1965.

same sense as that of its partner. The energy of the metastable triplet state may be used for bringing about a photochemical conversion, the long life favouring this possibility, or it may be dissipated as heat or as infra-red radiation (phosphorescence), or, if the molecule receives a second quantum, it may be converted to an excited triplet state. The second quantum need not be received by the same molecule but may be transferred from another in one of the ways already mentioned on p. 30. The energy of the excited triplet state would be expected to be available for photochemical conversion but this has not yet been demonstrated experimentally.

The energy of the absorbed photon, whether it is retained by the molecule which first absorbed it or transferred to another thus appears eventually as heat, fluorescence, phosphorescence or chemical energy. If the photochemical conversion is successfully carried out less energy can appear in the other forms. It is observed that living photosynthesizing algae remain cooler when illuminated than do similar dead ones and a calorimeter has actually been used to determine the efficiency of photosynthesis. The amount of fluorescence emitted by a plant is similarly an inverse measure of its efficiency in photosynthesis.

In the chlorophyll molecule the alternating double and single bonds provide orbitals which are continuous throughout the ring system so that an excited electron is not tied to a particular atom but moves about the whole molecule. Furthermore it is probable that the incorporation of the pigments in the chloroplast into a regular structure allows an even greater mobility of electrons. In a molecular crystal of this sort the electron orbitals overlap to such a degree that when excitation takes place the electron displaced is able to move freely about the whole structure. In this case the excited structure becomes electrically conducting although in its ground state it is non-conducting. It has been shown that dried chloroplasts can act as photoconductors in this way and there is some evidence that the entry into the conducting state depends on the metastable triplet state of chlorophyll. It is perhaps of significance that in photoconductors the "hole" left by the displaced electron is as mobile as the electron itself and that a "hole" left by a chemical conversion in one part of the structure may be filled by an electron displaced from quite another part. This again fits in with the idea of a photosynthetic unit in which a large number of chlorophyll molecules serve a single site at which the photochemical conversion takes place.

The return of electrons into holes may be delayed if they are caught by particular parts of the structure, such as breaks in the crystal continuity, which act as electron traps. A trapped electron may be restored into general circulation by thermal energy from the surroundings and then return to the ground state with emission of a light quantum. This evidently happens regularly in chloroplasts, for Strehler and Arnold, using sensitive photomultipliers, found in 1951 that photosynthetic tissues and cells are capable of delayed light emission. The emission lasts for a matter of minutes or even hours after the end of a light period and the wavelength is that of chlorophyll fluorescence. Were human eyes sufficiently sensitive we should see vegetation glow with deep red light after sunset. This luminescence is partly temperature dependent and appears to arise from the untrapping of electrons trapped in the chlorophyll complex.

Enhancement

Certain features of action spectra (p. 29) are difficult to account for on the supposition that chlorophyll *a* is the only photochemically active pigment in the chloroplast. If the absorption spectrum in Fig. 7 is compared with the quantum yield curve in Fig. 9a it will be seen that although the latter plunges abruptly downwards at wavelengths longer than 680 mμ, absorption by chlorophyll *a* is still appreciable down to 700 mμ. It seems from this that

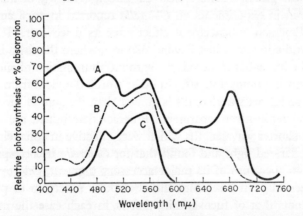

Figure 18 Absorption spectrum of the intact seaweed (A), action spectrum of photosynthesis (B), and absorption spectrum of the extracted phycoerythrin (C), for *Porphyra nereocystis*. (Redrawn from L. R. Blinks (1954), *Symp. Soc. gen. Microbiol.*, 4, 224.)

although the chlorophyll is absorbing quanta capable of producing excited states these are not used in photosynthesis. Even more curious is the action spectrum of the red seaweed, *Porphyra nereocystis* (Fig. 18), from which it appears that light absorbed by chlorophyll *a* in the red (600–700 mμ) and blue (400–480 mμ) regions is far less effective in photosynthesis than light absorbed by phycoerythrin from 520 to 560 mμ. Furthermore, it is found that light absorbed by phycoerythrin is more effective in exciting the fluorescence of chlorophyll *a* than is light absorbed by chlorophyll *a* itself. To explain this it was suggested that two forms of chlorophyll *a* exist in this alga, one highly fluorescent and capable of having energy transferred to it from phycoerythrin while the other is only weakly fluorescent.

The first evidence that different pigments may cause different photochemical reactions came from experiments in which another species of *Porphyra* was subjected to alternating periods of illumination with red and green light, the intensities of which were adjusted to give equal low rates of photosynthesis, without an intervening dark period (Blinks, 1957). If photosynthesis involves only one kind of photochemical process then all wavelengths of light should have the same qualitative effect and substitution of green for red and *vice versa* in this experiment should result in no changes in rate of photosynthesis. However, following each change to green there was a spurt in oxygen production followed by a decline, then a recovery to the average rate. This could be explained by effects on respiration or it might be that something produced during illumination with red light was promoting photosynthesis in the green. A more clear-cut effect, evidently of the same sort, was discovered in experiments on *Chlorella* reported in 1958 and has been named the Emerson enhancement effect after its discoverer. It was found that photosynthetic yields just beyond 680 mμ, where the drop in efficiency occurs, can be increased if there is background illumination of another wavelength efficient in photosynthesis. In other words the yield from a combination of long and short wavelengths is greater than the sum of the yields from the two wavelengths given separately. Emerson determined the action spectrum for the shorter wavelengths which were effective in enhancing the yield given by the far-red light and found that for *Chlorella* it corresponded with the absorption spectrum of its major accessory pigment, chlorophyll *b*. For the blue-green algae it resembles the absorption spectrum of phycocyanin and for diatoms that of fucoxanthin, that is, in each case the major accessory pigment is indicated. A proportion of chlorophyll *a* may also function like these accessory pigments, thus supporting the conclusion based upon fluorescence studies that was mentioned in the previous paragraph. The enhancement effect is observed even when the exposure to the two different wavelengths are separated by several seconds.

There are two possible explanations of the enhancement effect. One is that there is only one photochemical reaction in photosynthesis itself and this uses the bulk of the absorbed energy, a small amount of energy absorbed by another pigment system causing a subsidiary photochemical reaction which results in the production of an enzyme or intermediate which affects the rate of oxygen production. This hypothetical enzyme or intermediate could intervene in either photosynthesis or respiration for, as photosynthesis is being measured in terms of oxygen production, the observed effect could result from a decrease in respiration as well as an increase in photosynthesis. The respiration of the blue-green alga *Anacystis nidulans* has in fact been found to respond differently to green and red light and consequently, to obtain a true measure of enhancement, allowance must be made for these effects but, probably, they are usually rather small. Alternatively, there may be two photochemical reactions in photosynthesis proper, using comparable amounts of energy and one of them, at least, requiring the co-operation of the other if oxygen is to be evolved. Although the first possibility has not been entirely ruled out the second explanation has usually been accepted and used as a basis for speculation about mechanisms.

No enhancement effect could be found in *Rhodospirillum rubrum* by Blinks and van Niel (1963), but subsequent work has shown, nevertheless, that there are two distinct photochemical reactions in photosynthetic bacteria.

Chloroplast components other than photoactive pigments

Clues to the nature of the photochemical mechanism may be provided by study of the properties of the various substances found in chloroplasts, although, obviously, this might lead one astray since the fact that a substance is found in these organelles does not necessarily mean that it is directly concerned in this mechanism. Apart from chlorophylls, carotenoids and biliproteins those that seem to be of particular significance are the cytochromes, plastoquinones and ferredoxin.

The *cytochromes*, which are structurally related to chlorophyll in that they have a similar porphyrin ring system, but contain an atom of iron instead of magnesium, have long been known as electron carriers in aerobic respiration Several cytochromes are known to be located in the cell organelles specially concerned with respiration, the mitochondria (Plate IX). Cytochromes are also found in the chloroplasts, notably cytochrome *f* which is found only in chloroplasts in leaves and is present there in ten times the concentration of the others. Algal chloroplasts contain similar but not identical cytochromes. The respiratory cytochromes are accompanied by specific enzymes, cytochrome oxidases. No oxidase specific for cytochrome *f* is found in chloroplasts so that it is unlikely to be concerned in respiration. The idea that it may be involved in photosynthesis is strengthened by the finding that photosynthetic bacteria, which being anaerobic do not require respiratory cytochromes, nevertheless have special cytochromes in their chromatophores.

Another class of iron-containing electron-transporting compound found in chloroplasts is that of the *ferredoxins*. These substances are proteins and do not have the porphyrin ring structure of the cytochromes. They are also found in nitrogen-fixing bacteria and seem of special interest in connexion with photosynthesis because they have a lower oxidation–reduction potential than any other substance isolated from living organisms. The oxidation–reduction potential of a substance is a measure of its ability to carry out reductions, a substance being reduced by another substance only if this has a lower potential. Thus ferredoxin (-0.43 volt) is able to reduce NADP, a substance which itself possesses a very low oxidation–reduction potential (-0.321 volt). This takes place in the presence of a specific enzyme, a flavoprotein reductase, if a suitable source of reducing power such as illuminated chloroplasts is available. Spectrophotometric observations show that ferredoxin undergoes reduction in the presence of illuminated chloroplasts. Its reduced form is stable under anaerobic conditions but it reverts to the oxidized form in the dark if NAPD is added.

FERREDOXIN REDUCED⟍⟋FLAVOPROTEIN OXIDIZED⟍⟋NADPH$_2$
FERREDOXIN OXIDIZED⟋⟍FLAVOPROTEIN REDUCED⟋⟍NADP

Ferredoxin occurs in the ratio of one molecule to 400 of chlorophyll, the same ratio as that of cytochrome *f* to chlorophyll, which suggests a key rôle in the photosynthetic unit.

A blue, copper-containing protein, *plastocyanin*, has also been found in chloroplasts and appears to act as an electron-transferring catalyst.

A great variety of quinoidal substances having the general formula:

$$
\begin{array}{c}
\text{O} \\
\parallel \\
\text{C} \\
R_1\text{—C}\diagup \diagdown\text{C—}R_3 \qquad \text{C} \\
R_2\text{—C}\diagdown \diagup\text{C—[C—C}=\text{C—C]}_n \\
\text{C} \\
\parallel \\
\text{O}
\end{array}
$$

are found in chloroplasts in amounts of the order of one-tenth or more of those of chlorophyll. Vitamin K is one such substance but *plastoquinone* (for which $R_1 = CH_3$, $R_2 = CH_3$, $R_3 = H$ and $n = 9$ in the above formula) is the most abundant and is characteristic of plants. Quinones can act as electron carriers and, as we shall see, there is evidence that they undergo oxidation–reduction during photosynthesis.

Changes in absorption spectra in illuminated chloroplasts

The absorption spectrum of a substance changes when it becomes excited by absorption of a light quantum or when it undergoes a chemical reaction such as oxidation–reduction. A possible approach to the study of the photochemical reactions is therefore to observe the alterations in the absorption spectra of chloroplasts which occur under various conditions. This presents technical difficulties, because the changes are small and occur in a densely absorbing medium, but these can be overcome by employing a compensating beam of light, which passes through a control suspension, and measuring the differences in absorption between this and the analysing beam, which passes through the experimental suspension. Using such an arrangement one can then observe either the transient effects following a flash of light or the differences between one steady state and another.

Changes having a half-life of about 5×10^{-5} sec following flash illumination indicate the formation of metastable states of chlorophyll in the chloroplast, but the rôle of these in photosynthesis remains uncertain. Many investigators using a variety of green cells and tissues have observed absorption changes which seem to correspond to the oxidation and reduction of cytochromes. Thus in *Chlorella* suspensions a minimum has been recorded in the difference spectrum at 420 mμ which appears to correspond with the oxidation of cytochrome in the light and its reduction in the dark. With the red alga *Porphyridium cruentum* (Plate XI), in which the absorption bands of cytochrome f are not so much obscured by chlorophyll, clear-cut evidence of the photoxidation of this cytochrome has been obtained. Evidence obtained using photosynthetic bacteria is, however, more direct. Oxygen-independent photo-oxidation of cytochrome C has been observed in *Rhodospirillum rubrum* (Fig. 19) and with chromatophore preparations from the same organism the cytochrome became oxidized when photophosphorylation was taking place although it did not when this was prevented by absence of phosphate acceptor (ADP). It therefore seems highly probable that the cytochromes are directly involved in the photochemical reactions.

In algae and higher plants a decrease in absorption occurs near the red absorption peak of chlorophyll upon illumination. The substance responsible for this has been designated as P 700 because the decrease is greatest at 700 mμ. It is probable that it is a modification of chlorophyll a and it occurs in about the same molar concentration as cytochrome f, that is, one molecule for every 400 chlorophyll molecules. Its spectral properties may be explained if it is chlorophyll a in a special structural environment and coupled with cytochrome f. Since its absorption band overlaps with those of the other chlorophylls and is on the longer wavelength side of them, it should be effective in collecting energy from them. The decrease in absorption, which results from oxidation, occurs during illumination with far-red light and is reversed by shorter wavelength light. These facts suggest that P 700 is associated with the active centre of the photosynthetic unit and takes part in electron transport together with cytochrome *f*. Flash experiments indicate that cytochrome *f* is the immediate reductant for P 700.

Corresponding with P 700 in green plants the purple bacterium *Rhodopseudomonas spheroides* possesses a modification of bacteriochlorophyll characterized by bleaching in the 870 mμ absorption region on illumination

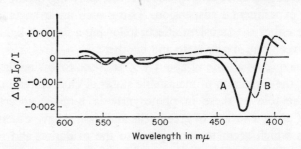

Figure 19 Changes in absorption spectrum induced by illumination in the photosynthetic bacterium *Rhodospirillum rubrum*. A, difference between the absorption spectra of cells in the light and in the dark, showing a maximum at 425 mμ; B, the difference between the absorption spectra of oxidized and reduced cytochrome C. (Redrawn from L. N. M. Duysens (1957) in *Research in Photosynthesis*, Interscience, 164.)

and hence termed P 870. This bleaching is likewise due to oxidation and can be brought about by chemical oxidation as well as by light. P 870 seems to be associated with the photochemical reaction centres, since it is absent in a mutant of *R. spheroides* which is incapable of photosynthesis although possessing the normal complement of light-harvesting bacteriochlorophyll and carotenoids. In fresh cells of this purple bacterium there is one molecule of P 870 for every forty molecules of ordinary bacteriochlorophyll. Impressive demonstration that the bleaching in light is a purely photochemical reaction is given by the finding that it occurs at temperatures approaching absolute zero, at which chemical reactions are of course completely stopped. P 870 in *R. spheroides*, its counterpart P 890 in *Rhodospirillum rubrum* and *Chromatium*, and P 700 in green plants thus appear to be the photochemical reaction centres of the photosynthetic unit. It may be, however, that their light reactions are due to photoxidations which have only a secondary relationship to photosynthesis.

A relatively new technique which may well give considerable information about these and other events is that of electron spin resonance (ESR). Because of their spin, electrons behave like tiny bar magnets, but since these spins are normally paired off against each other this is not apparent and matter is not usually magnetic. In certain situations, as in free radicals and triplet states, electrons become unpaired and the material exhibits paramagnetic properties, showing characteristic resonance absorption of energy in a magnetic field, which can be detected by the electron spin resonance spectrometer. As would be expected, such a resonance is observed in photosynthesizing cells and chloroplasts and changes with illumination. There is some evidence that a rapidly decaying resonance called the R-signal is associated with chlorophyll and in particular with P 700, or, in purple bacteria with P 870. Thus a mutant of the green alga *Chlamydomonas* which lacks P 700 fails to show this signal.

Cyclic and non-cyclic photophosphorylation as electron-flow processes

So far we have considered the initial stages of the conversion of light energy in photosynthesis. The problem of what happens can also be tackled from the other end by biochemical investigation of the end-products.

As we saw in the first chapter, isolated chloroplasts are capable of two main photochemical processes, cyclic and non-cyclic photophosphorylation. In the former, ATP is produced at the expense of absorbed light energy. However, ATP may also be produced by purely chemical "dark" reactions in respiration and this is known to be coupled with the fall of an electron from a higher to a lower energy level, the free energy thus released being used in the synthesis of the high-energy phosphate bond. This reaction requires a substrate (the electron donor) and oxygen (the electron acceptor). By analogy one might expect the synthesis of ATP in cyclic photophosphorylation to be coupled with electron transport but, since an electron donor is not required nor oxygen consumed, it must be a closed circuit with light energy being used to push electrons to a higher energy level whence they return to their original starting point. Probable electron carriers for the return, "downhill", journey are vitamin K or plastoquinone and cytochrome. From their oxidation-reduction potentials it appears that plastoquinone or cytochrome B_6 and cytochrome *f* might provide for a two-stage return, both of which might be coupled with ATP synthesis, the cytochrome *f* step being the second, lower energy level stage, since this substance has a higher oxidation reduction potential than the other two (see p. 48). Some evidence for a two-stage process comes from investigation of the effects of phenazine methosulphate, an artificial electron-transporting substance, on cyclic photophosphorylation. Addition of this substance speeds up the process, evidently by short-circuiting one stage so that the yield of ATP is halved if light is limiting. On this basis Arnon has proposed the scheme for cyclic phosphorylation which is summed up in Fig. 20.

This electron-flow theory has been extended by Arnon and his collaborators to cover non-cyclic photophosphorylation. Here ATP formation is coupled with the reduction of NADP to $NADPH_2$ and hence a source of hydrogen, or, what is equivalent, a source of electrons, is required. In green plants the ultimate hydrogen donor is normally water and the oxygen evolution is

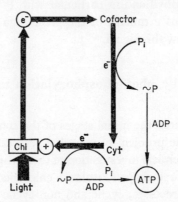

Figure 20   Scheme for electron transfers in anaerobic cyclic photophosphorylation. (Redrawn from F. R. Whatley and M. Losada, in *Photophysiology*, **1**, 111; Academic Press.)

interpreted as a means of disposing of the oxidized portion of the water molecule left after the hydrogen has been abstracted. Other hydrogen donors may be substituted for water. Thus chloroplasts that have lost the capacity to evolve oxygen, either through ageing or inhibition with the herbicide CMU (*p*-chlorophenyl dimethylurea) or DCMU (dichlorophenyl-dimethyl-urea), continue to reduce NADP if ascorbic acid is provided as a hydrogen source with the dye 2,6-dichlorophenolindophenol (DCPIP) to act as a hydrogen carrier. Furthermore, using such dyes, it has proved possible to separate non-cyclic photophosphorylation into two separate photochemical reactions. If the dye is supplied in the oxidized form (A) to isolated chloroplasts a Hill reaction occurs and oxygen is evolved. This can be envisaged as:

16) $$2OH^- + 2A \xrightarrow{\text{light}} \tfrac{1}{2}O_2 + 2A^- + H_2O$$

OH^- being supplied by the ionization of water and the hydrogen ion left balancing the A^- produced. If the dye is supplied in the reduced form (A^-) to CMU — inhibited chloroplasts it replaces water as a source of electrons and we get photophosphorylation and $NADPH_2$ production without oxygen evolution:

17) $2A^- + NADP + 2H^+ + ADP + P_i \xrightarrow{\text{light}} 2A + NADPH_2 + ATP$

These two processes may be represented in terms of electron flow as in Figs. 21 and 22. The formation of ATP is supposed to involve both chlorophyll

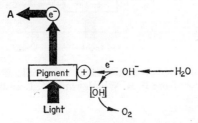

Figure 21 Scheme for electron transfers occurring in illuminated chloroplasts in the presence of a Hill reagent, A. (Redrawn from F. R. Whatley and M. Losada, in *Photophysiology*, **1, 111**; Academic Press.)

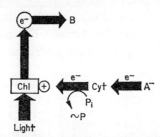

Figure 22 Scheme for electron transfers in non-cyclic photophosphorylation by chloroplasts in the presence of a source of electrons, A^-. B is an electron acceptor. This resembles non-cyclic photophosphorylation as it occurs in bacteria. (Redrawn from F. R. Whatley and M. Losada, in *Photophysiology*, **1, 111**; Academic Press.)

and cytochrome as in cyclic photophosphorylation, but an accessory pigment is probably concerned in the second photochemical reaction as we shall see later (p. 56). The two processes add up to give equation 12 for non-cyclic photophosphorylation given on page 10 and can be put together in the scheme given in Fig. 23. The indophenol dye is an artificial hydrogen carrier substituting for a natural intermediate which has not been identified with certainty but which is possibly plastoquinone. Evidence pointing to this is that chloroplasts lose the ability to carry out the Hill reaction and evolve oxygen after they have been extracted in the freeze-dried condition with petroleum ether, but their activity is restored if plastoquinone is supplied. Chloroplasts

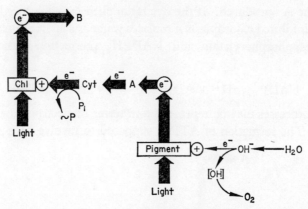

Figure 23 Scheme for electron transfers in non-cyclic photophosphorylation in chloroplasts. (Redrawn from F. R. Whatley and M. Losada, in *Photophysiology*, **1**, 111; Academic Press.)

from which plastoquinone has been extracted can still effect the photochemical reduction of NADP and production of ATP if ascorbic acid and DCPIP are supplied as a source of electrons (Fig. 24).

Bacterial chromatophores resemble chloroplasts in being able to carry out both cyclic and non-cyclic phosphorylation. The purple sulphur bacterium *Chromatium* is capable of using molecular hydrogen directly for reduction of

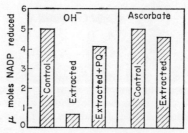

Figure 24 Effect of plastoquinone on the photoreduction of NADP. The "extracted" chloroplasts were extracted with heptane. Plastoquinone (PQ) was added in the ratio of 0·05 mg/0·5 mg chlorophyll. In the "ascorbate" system ascorbate + dye acted as the electron donor and the reactions were carried out in the presence of the inhibitor CMU. (Redrawn from F. R. Whatley and A. A. Horton (1963), *Acta. chem. Scand.*, **17**, 135.)

NAD (which has the same rôle as NADP in green plants), then only requiring light for cyclic photophosphorylation in order to carry out synthesis of carbohydrate. However, photosynthetic bacteria also use electron donors, for example thiosulphate or succinate, which do not have sufficiently low oxidation–reduction potentials for the direct reduction of NAD. An extra energy supply is needed in this case and this may be provided by light via an electron-flow mechanism similar to that postulated as operating in green plants and algae. Electrons of moderate reducing potential are transferred via cytochrome to chlorophyll which, given a quantum of light energy, raises them to a reducing potential greater than that of NAD. Besides being used for this purpose, these electrons can be used in *Chromatium* for the production of molecular hydrogen from hydrogen ion or, perhaps, for the reduction of molecular nitrogen to ammonia, that is to say for the photosynthetic fixation of atmospheric nitrogen (see p. 77).

The chemical effects of the two light reactions

Two light reactions are required to give maximum yields in photosynthesis by green plants and algae, and, as we have just seen, non-cyclic phosphorylation can be separated into two photochemical reactions. The hope which naturally arises that these pairs correspond seems to be justified. The reactions represented by equations 16 and 17 above show differing sensitivities to different wavelengths (Fig. 25). The normal reaction resulting in the

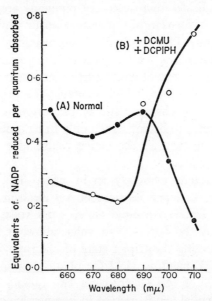

Figure 25 The quantum yield for photoreduction of NADP in relation to wavelength. A, with normal chloroplasts, in which both light reactions function; B, with chloroplasts treated with the inhibitor DCMU and using DCPIPH as hydrogen donor, when only one light reaction functions. (Redrawn from G. E. Hoch (1965) in *Biochemical Dimensions of Photosynthesis*, p. 5; Wayne State University Press.)

reduction of NADP, which requires both photochemical reactions, shows a decline in quantum yield at far-red wavelengths whereas the photoreduction in the presence of DCMU and reduced DCPIP, which does not involve oxygen production and presumably depends on only one photochemical reaction, shows a steep rise in this region. It seems then that the long-wavelength photochemical reaction (system I) is responsible for reduction of NADP and the short-wavelength system II for oxygen production. System I may be observed separately from the other in chloroplasts illuminated with long-wavelength light or treated with DCMU, but system II is not so easily studied in isolation. However, a *Scenedesmus* mutant, obtained by Bishop, seems to lack system I and cannot photoreduce NADP.

Evidence already mentioned indicates the nature of the pigments concerned in these two reactions. The form of chlorophyll *a* known as P 700 becomes oxidized in long wavelengths and reduced in shorter wavelengths. It thus is likely to be the photoactive pigment in system I, which donates the electrons, thereby becoming oxidized, for the reduction of ferredoxin which in turn reduces NADP. The weakly fluorescent chlorophyll *a* in red algae (p. 46) is also associated with system I. The action spectrum for the shorter wavelength effect indicates that accessory pigments are mainly concerned in this—chlorophyll *b*, fucoxanthin, phycocyanin or phycoerythrin according to the kind of plant. This conclusion is borne out by other experiments. By treatment of chloroplasts with detergent it is possible to separate two kinds of protein complex, one of which carries out reactions characteristic of photosystem I and the other, which contains the greater proportion of accessory pigment, those characteristic of photosystem II. However, the accessory pigment seems to be concerned only in light absorption and, as in photosystem I, the pigment in the active centre of photosystem II is a chlorophyll *a* modification.

These two photosystems evidently act in series. The energy of a quantum of red light (40 kcal per Einstein or 1·7 electron volts) is sufficient to raise an electron from its ground state, +0·8 volts, to the oxidation-reduction potential of NADP, −0·32 volts, allowing for losses such as those involved in forming the triplet state of chlorophyll. However, since the generally accepted quantum yield for photosynthesis (see p. 6) corresponds to the transfer of one electron per two quanta it is reasonable to suppose that the potential gap is crossed in two nearly equal stages. In this case there must be one or more electron carriers with intermediate oxidation-reduction potentials. Plastoquinone and cytochrome *f* meet this requirement and since they react in one direction when system I is activated by light and in the other direction when system II is activated it seems that either they are situated in between the two systems or react with some intermediate which is. Thus plastoquinone shows by changes in its absorption spectrum that it is reduced when illuminated with short wavelengths and oxidized with long wavelengths, that is, it accepts electrons from system II and hands them on to system I. Similarly cytochrome f responds by becoming reduced in short wavelengths and oxidized in long wavelengths (Fig. 26).

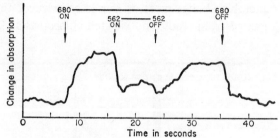

Figure 26 The time course of cytochrome oxidation in the red alga *Porphyridium cruentum* when illuminated with light of wavelengths 680 and 562 mμ. The curve shows that the cytochrome is oxidized in light of 680 mμ but reduced in light of 562 mμ. (Redrawn from L. N. M. Duysens (1963), *Proc. Roy. Soc. B,* **157,** 301.)

Some comment should be made about the oxygen-liberating mechanism associated with system II. Very little is known about this. The old idea that hydrogen peroxide was formed from the [OH] residues and decomposed by the enzyme catalase

$$18) \qquad 2[OH] \rightarrow H_2O_2 \xrightarrow{\text{catalase}} H_2O + \tfrac{1}{2}O_2$$

does not seem to be correct since catalase in living algae, *Scenedesmus,* can be inhibited by cyanide without affecting the evolution of oxygen. Possibly cytochromes are involved, but the only definitely established requirements are for chloride ion and traces of manganese. In the absence of chloride, chloroplasts behave like bacterial chromatophores, being able to carry out anaerobic cyclic photophosphorylation but not being able to evolve oxygen. Algae deficient in manganese show a low rate of photosynthesis but recover in less than one hour if a small amount of manganous sulphate is supplied. Photoreduction by these same algae (see p. 7) and bacterial photosynthesis, neither of which involves oxygen liberation, are not affected by manganese deficiency.

The sequence of reactions in photosynthesis

The evidence that has been presented above can be fitted into a scheme such as that given in Fig. 27. A more compact linear version of the same scheme is:

$$NADP \leftarrow Fd \leftarrow \text{system I (P 700)} \leftarrow Cyt \leftarrow PQ \leftarrow \text{system 2} \leftarrow H_2O$$

where Fd = ferredoxin, Cyt = cytochrome f and PQ = plastoquinone, and the electron flow is visualized as taking place from right to left leaving the oxidizing portion of the water molecule to be disposed of on the right. It should be noted that components besides those indicated in Fig. 27 are certainly involved but have been omitted so as not to complicate the picture unduly.

It must be remembered that most of our knowledge of these chemical events has been gained from studies on isolated chloroplasts. These do not

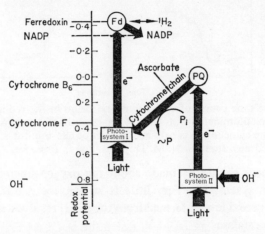

Figure 27 Scheme for electron transfer in non-cyclic photophosphorylation in terms of redox potentials. (Redrawn from F. R. Whatley and M. Losada, in *Photophysiology*, **1, 111**; Academic Press.)

necessarily function in the same way as chloroplasts in intact cells and, indeed, the fact that the chloroplast preparations with which most of the work which has been described was done were much less active than a corresponding amount of living photosynthetic material shows that they were impaired in some way. Only recently have chloroplast preparations been obtained having as much as two-thirds of the activity of chloroplasts in the living cell. Furthermore there are discrepancies between their theoretical and actual activities that need resolving. The theoretical yield of ATP in non-cyclic phosphorylation by chloroplasts is 1 per $NADPH_2$ produced but for the reduction of a molecule of carbon dioxide to the level of carbohydrate 2 molecules of $NADPH_2$ and 3 of ATP are required (p. 66) and if the ATP requirements for the conversion of carbohydrate to cell material are taken into account the latter figure must be raised to 6. According to the scheme given in Fig. 27 8 quanta would produce 2 $NADPH_2$ and 2 ATP. The shortage of ATP might be made up by cycling of system I but this would consume quanta over and above the 8 or less that is actually found to be required. Experimentally, however, yields of ATP greater than 1 per $NADPH_2$ and sufficient to provide for carbon dioxide assimilation have been sometimes observed in non-cyclic photophosphorylation. Another approach is that of Arnon (1967), who has reported evidence which suggests that cyclic and non-cyclic photophosphorylations involve separate photochemical events. On his hypothesis complete photosynthesis depends not on the joint contribution of two photochemical steps acting in series to achieve non-cyclic photophosphorylation but on the joint contribution of two distinct photochemical systems, cyclic and non-cyclic photophosphorylation, acting in parallel to supply the requirements for the reduction of carbon dioxide.

There has been difficulty in relating bacterial photosynthesis to the scheme given in Fig. 27. Here, except in the matter of oxygen evolution, there are so many resemblances to green plant photosynthesis—for instance the chlorophylls and cytochromes are closely related, photophosphorylation occurs in both groups and the enzyme systems for carbon dioxide assimilation are similar—that we must accept the two processes as having similar basic mechanisms. Furthermore, as we have seen (p. 7), algae can be made to carry out something closely resembling bacterial photosynthesis (photoreduction), by adaptation under anaerobic conditions. The absence of an enhancement effect in bacteria led to the idea that they possess only photosystem I. If this were so then the quantum requirement should be one per electron, that is, four for every two NAD molecules reduced to $NADH_2$. Determinations of the quantum requirement in bacterial photosynthesis has shown, however, that it is about two per electron regardless of whether the electron donor is hydrogen which can reduce NAD directly or thiosulphate which supplies electrons at a low potential. The discovery that the bacteria contain two distinct photosystems—one a cyclic system generating ATP and the other coupled to the oxidation of the hydrogen donor (Sybesma, 1969)—seems to have resolved this difficulty.

Still much remains to be learnt about the photochemical process. The electron flow hypothesis seems to account adequately for most of the known facts, but Franck put forward modifications of van Niel's theory (that the photolysis of water is the key act in photosynthesis) which take account of the existence of two light reactions. However, different though the currently accepted electron-flow hypothesis seems from that of van Niel, they are essentially similar in that they both envisage the primary photochemical act as the utilization of light energy via chlorophyll to separate oxidizing and reducing entities, and it may be as well to conclude this chapter with a simple diagram which emphasizes this point (Fig. 28).

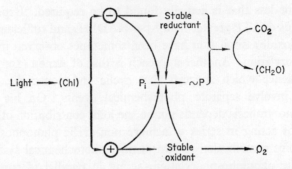

Figure 28 A general scheme of the mechanism of photosynthesis in terms of the production of an oxidant, $\oplus$, and reductant $\ominus$. (Redrawn from R. K. Clayton (1965), *Molecular Physics in Photosynthesis*, Blaisdell Publishing Co.)

Further Reading

Arnon, D. I. (1966) The photosynthetic energy conversion process in isolated chloroplasts. *Experientia*, **22**, 1–15.

Arnon, D. I., Tsujimoto, H. Y., and McSwain, B. D. (1967) Ferredoxin and photosynthetic phosphorylation. *Nature*, **214**, 562–6.

Bishop, N. I. (1966) Partial reactions of photosynthesis and photoreduction. *Ann. Rev. Plant Physiol.*, **17**, 185–208.

Boardman, N. K. (1970) Physical separation of the photosynthetic photochemical systems. *Ann. Rev. Plant Physiol.*, **21**, 115–40.

Clayton, R. K. (1965) *Molecular Physics in Photosynthesis*. Blaisdell Publishing Co., New York, Toronto and London.

Gest, H., San Pietro, A., and Vernon, L. P., editors (1963) *Bacterial Photosynthesis*. Antioch Press, Yellow Springs, Ohio.

Goodwin, T. W., editor (1966, 1967) *Biochemistry of Chloroplasts*, vols. I and II. Academic Press, London and New York.

Kamen, M. D. (1963) *Primary Processes in Photosynthesis*. Academic Press, New York and London.

Rabinowitch, E., and Govindjee (1969) *Photosynthesis*. John Wiley & Sons Inc., New York and London.

Shibita, K., Takamiya, A., Jagendorf, A. T., and Fuller, R. C. (Editors) (1968) *Comparative Biochemistry and Biophysics of Photosynthesis*. University of Tokyo Press and University Park Press.

Whatley, F. R., and Losada, M. (1964) The photochemical reactions of photosynthesis. In *Photophysiology* edited by A. C. Giese, vol. I, pp. 111–54. Academic Press, New York and London.

5 *The path of carbon assimilation*

AS outlined in Chapter 1, the principal path of photosynthetic carbon assimilation is thought to be one in which carbon dioxide is brought into combination by reacting with a 5-carbon acceptor substance, ribulose-1,5-diphosphate (RuDP), this immediately splitting to give two molecules of the 3-carbon compound phosphoglyceric acid (PGA), the first stable product. PGA is then reduced to a triose sugar by means of the assimilatory power generated by the photochemical reactions and from this the acceptor RuDP is re-formed. We must now consider this cycle in more detail.

Dark fixation of carbon dioxide

Carbon dioxide fixation occurs in all living organisms although in non-photosynthetic forms it is normally masked by the output of carbon dioxide in respiration. This "dark" fixation was first detected in bacteria by Wood and Werkman (1935), and similar processes were later found in animal tissues, fungi and green plants kept in the dark. The reactions are of this type:

19) $CH_2:CO\sim P.COOH + {}^{14}CO_2 \rightarrow HOO^{14}C.CH_2CO.COOH + P_i$

carbon dioxide being added on as a carboxyl (—COOH) group to a suitable organic compound, a process known as carboxylation. If carbon dioxide labelled with radiocarbon is used, as indicated in the equation, the occurrence of carboxylation is readily demonstrated and the position of the incorporated carbon in the product may be ascertained. In this example phospho-enol-pyruvic acid gives oxaloacetic acid, the reaction being catalysed by phospho-enolpyruvic carboxylase. Since reduction of carbon dioxide is not involved, the energy requirement of carboxylation is small and in this case, but not necessarily in others, it is provided by the organic phosphate group and no outside source of energy is required. Many of the decarboxylations occurring in respiration are reversible and consequently, if a plant is supplied with ^{14}C-carbon dioxide in the dark, a chromatogram made of an extract of its tissues will show several radioactive spots corresponding to the organic acids which are intermediates in this process.

The formation of PGA in photosynthetic carbon fixation

If a similar experiment to that just described is carried out with photosynthetic cells or tissues in the light the chromatogram obtained shows a different pattern. Because of the rapid transfer of radiocarbon to a wide variety of compounds the period of exposure to ^{14}C-carbon dioxide has to be short if useful results are to be obtained and some indication of the procedure should be given here. Suspensions of the unicellular algae *Chlorella* or *Scenedesmus* are usually used, because of the ease with which they may be handled, and are first allowed to reach a steady state of photosynthesis in the presence of ordinary carbon dioxide. ^{14}C-carbon dioxide is supplied by injecting ^{14}C-bicarbonate into the rapidly stirred suspensions and then, after a suitable period of further photosynthesis, the suspension is run into boiling alcohol. This stops all enzyme activity and extracts water-soluble substances from the cells. The substances contained in the filtered extract are separated on a two-dimensional chromatogram. On the chromatogram will be spots corresponding to both unlabelled and radioactive substances and the latter may be located either with a Geiger–Müller counter or by leaving the paper in contact with a photographic film, which will develop, when processed, a black spot corresponding exactly in position and shape to each patch of radioactivity on the paper. Particular substances may be identified by their position on the chromatogram, by chemical tests, or by extracting them and comparing their chromatographic behaviour with that of authentic samples.

It will be seen from Plate II*b* that, in a chromatogram made after an exposure in the light, some of the acids which would be heavily labelled as a result of dark fixation of $^{14}CO_2$, succinic and glutamic, for example, are now scarcely radioactive. On the other hand, numerous substances, notably phosphoglyceric acid (PGA) and sugar phosphates, which do not become appreciably labelled in the dark, are heavily labelled after a light exposure of a minute or so. A few substances, the amino-acid alanine and malic acid, for instance, become labelled to more or less the same extent in both light and dark. At one time it was thought that malic acid might be an important intermediate in the photosynthetic fixation of carbon dioxide, but the fact that its formation may be inhibited without affecting fixation in other products disproved this idea. As the period of exposure to light in the presence of ^{14}C-carbon dioxide is shortened PGA becomes the most conspicuously labelled substance and if the period is 5 seconds or less as much as 80 per cent of the ^{14}C fixed is found in this substance. It was thus established that PGA is the first stable product of photosynthetic carbon assimilation and careful search failed to reveal any unstable products preceding it.

By chemical breakdown of the labelled PGA and examination of the fragments it is possible to determine the average extents to which the three carbon atoms in the molecule are radioactive. After a few seconds' exposure only the carbon of the carboxyl group is labelled, that is to say we have

$$CH_2O—P.CHOH.{}^{14}COOH,$$

but after two minutes' exposure to radioactive carbon dioxide the three carbon atoms are nearly equally labelled, thus, ${}^{14}CH_2O—P.{}^{14}CHOH.{}^{14}COOH$. This suggests that PGA is formed by addition of carbon dioxide, by a carboxylation reaction similar to those which are responsible for dark fixation, to an acceptor substance which itself is rapidly reformed from the carbon fixed. The first thought was that the acceptor must have a molecule containing two carbon atoms but extensive investigations failed to implicate any such substance in PGA formation. The same result could, however, be produced if carbon dioxide were added on to a larger molecule, the product of this reaction subsequently splitting to give two fragments one or both of which are PGA. A 5-carbon sugar phosphate, ribulose-1,5-diphosphate (RuDP), which was known to be among the substances rapidly becoming radioactive during

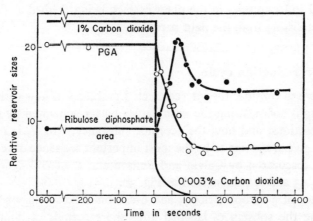

Figure 29 Effect of lowering of carbon dioxide concentration on the reservoir sizes of phosphoglyceric acid (PGA) and ribulose diphosphate. (Redrawn from M. Calvin and J. A. Bassham (1962), *The Photosynthesis of Carbon Compounds*, W. A. Benjamin, Inc.)

photosynthesis in the presence of radioactive carbon dioxide, was, in fact, shown to be the acceptor. Some of the evidence for this is shown in Fig. 29. *Scenedesmus* was kept under steady state conditions of photosynthesis with radioactive carbon dioxide until successive samplings showed the labelling in PGA, RuDP and other substances immediately involved in photosynthesis to have reached a constant level. The carbon dioxide concentration was then quickly lowered from 1 per cent to 0·003 per cent, all other conditions, including the specific radioactivity of the carbon dioxide, being kept constant. The size of the PGA reservoir thereupon fell whereas that of RuDP rose, at least initially, as one would expect if the reduced availability of carbon dioxide slowed down the conversion of RuDP to PGA. Changes in the reservoir sizes of these two substances on transfer of the alga from light to dark were also found to be consistent with the idea that RuDP is the acceptor substance.

RuDP reacts with carbon dioxide giving a highly unstable 6-carbon compound which *in vitro* splits to give two molecules of PGA.

20) $$CO_2 + RuDP + H_2O \rightarrow 2\,PGA$$

It seems possible, however, that in the living plant the 6-carbon compound may behave otherwise (see p. 67). The reaction, which is catalysed by ribulose diphosphate carboxylase, proceeds spontaneously and, apparently, irreversibly. It requires no supply of energy or source of hydrogen and is thus not directly dependent on the photochemical reactions. This can be demonstrated by supplying a plant, which has been photosynthesizing with ordinary carbon dioxide, with $^{14}CO_2$ at the same moment that the light is cut off. ^{14}C-labelled PGA appears in the plant extract, being formed at the expense of RuDP persisting from the light period.

The carbon reduction cycle

We now have to consider the interconnected problems of how carbon reduction is brought about, using the assimilatory power generated by the photochemical reactions, and how the acceptor substance, RuDP, is regenerated from PGA. What appears to be the most important sequence of reactions involved was discovered by Calvin and Benson and their colleagues between 1946 and 1953 and is summarized in Fig. 30.

Phosphates of 3-carbon (triose), and 6-carbon (hexose) sugars are prominent among the substances labelled after short periods of photosynthesis with $^{14}CO_2$. Now, both PGA and triose phosphate are known to be intermediates in the breakdown of hexose phosphate in respiration, in which ATP and $NADPH_2$ (or $NADH_2$) are produced, and it seemed a reasonable working

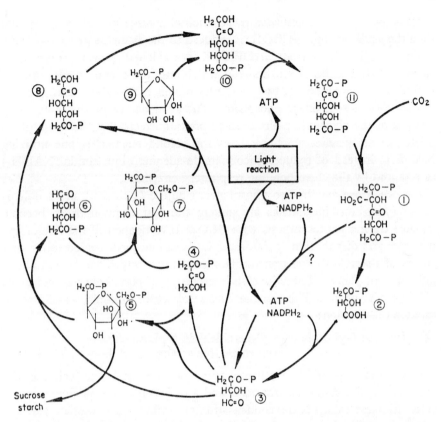

Figure 30 The photosynthetic carbon reduction cycle. (1) 2-carboxy-3-keto-1,5-diphosphoribitol, (2) 3-phosphoglyceric acid (PGA), (3) glyceraldehyde-3-phosphate, (4) dihydroxyacetone phosphate, (5) fructose-1,6-diphosphate, (6) erythrose-4-phosphate, (7) sedoheptulose-1,7-diphosphate, (8) xylulose-5-phosphate, (9) ribose-5-phosphate, (10) ribulose-5-phosphate, (11) ribulose-1,5-diphosphate (RuDP). (Modified from M. Calvin and J. A. Bassham (1962), *The Photosynthesis of Carbon Compounds*, W. A. Benjamin, Inc.)

hypothesis that sugars might be formed in photosynthesis by a reversal of this process, ATP and NADPH$_2$ being supplied. The order of appearance and positioning of tracer carbon in triose and hexose phosphates was found to be in approximate agreement with this idea and it is now accepted that hexose phosphate can be produced by the following reactions:

21) $\qquad$ PGA + ATP $\rightarrow$ diphosphoglyceric acid

22) diphosphoglyceric acid + NADPH$_2$ $\rightarrow$ glyceraldehyde phosphate
$\qquad\qquad\qquad\qquad\qquad\qquad\qquad\qquad$ + NADP + P$_i$

23) $\qquad$ 2 glyceraldehyde phosphate $\rightarrow$ hexose phosphate + P$_i$

That is to say, PGA is activated by addition of another phosphate group and then the acidic group, —COOH, is reduced to the aldehyde group, —CHO, by means of hydrogen from $NADPH_2$. Glyceraldehyde phosphate is a triose sugar and by a series of reactions, of which only the overall result is represented in the equation 23, two molecules of this are condensed to give one molecule of hexose sugar phosphate. The sequence of events is an exact reversal of the glycolytic pathway in respiration except that the triose phosphate dehydrogenase which catalyses the second reaction is one utilizing $NADPH_2$ instead of reduced nicotinamide-adenine dinucleotide ($NADH_2$) as required by the corresponding respiratory enzyme.

Besides triose and hexose phosphates, ribulose (5-carbon) and sedoheptulose (7-carbon) phosphates are among the first compounds to become labelled during photosynthesis with $^{14}CO_2$. Investigation of the rate of appearance and distribution of the tracer in these intermediates led to the discovery of a series of sugar transformations of a type which had hitherto been unknown in plants. These involve interchange of parts between molecules of sugar phosphates. The enzyme transketolase, for example, was found to catalyse the reaction:

24) fructose-6-phosphate + glyceraldehyde-3-phosphate →
            erythrose-4-phosphate + xylulose-5-phosphate

that is to say, a 4-carbon (erythrose) and a 5-carbon molecule (xylulose) are produced from a 6-carbon (fructose) and a 3-carbon (glyceraldehyde) molecule. Aldolase catalyses the condensation of erythrose-4-phosphate and another triose, dihydroxyacetone phosphate, to give sedoheptulose diphosphate and this, after loss of a phosphate group, will interchange with glyceraldehyde-3-phosphate in the presence of transketolase to give two different pentoses, ribose-5-phosphate and xylulose-5-phosphate. Both of these pentoses can be converted to ribulose-5-phosphate by the enzymes isomerase and epimerase respectively. From ribulose-5-phosphate, RuDP (ribulose-1,5-diphosphate), the carbon dioxide acceptor, is produced by the action of phosphoribulokinase with ATP donating the second phosphate group. Together with those synthesizing hexose from triose these enzymes thus form a highly complex and flexible system into which sugar can be fed as triose and withdrawn as the pentose carbon dioxide acceptor, hexose or other form according to demand.

In the steady state when hexose sugar is the end-product the overall reaction can be summarized as

25) $C_1 + C_5 \rightarrow 2C_3 \rightarrow C_6 \rightarrow C_5 + \frac{1}{6}$ hexose

this consuming the hydrogen from two molecules of $NADPH_2$ and the energy from three ATP molecules. One complete turn of the cycle, in the course of which every component reaction represented in Fig. 30 must occur at least once, requires the input of three carbon dioxide molecules.

The photosynthetic carbon reduction cycle as outlined above is generally accepted as substantially correct. The observed patterns of distribution of tracer carbon have not been explained satisfactorily on any other basis and most of the enzymes concerned have been shown to be present in photo-synthetic tissues. This is not to say that there are no unexplained discrepancies or that it represents the complete system concerned in carbon dioxide fixation.

One fact not accounted for by the cycle in its simplest form is that following photosynthesis with $^{14}CO_2$ the labelling in glucose is sometimes not symmetrical. If hexose is indeed formed by the condensation of two triose mole-cules both derived from PGA then the extent of labelling in corresponding carbon atoms should be the same, but distinct differences have been observed by several workers. Several explanations for this are possible, one being that in photosynthesis in the plant PGA may be bypassed by direct reduction to a sugar of the unstable compound of RuDP and carbon dioxide. Another possi-bility is that some of the PGA is bound to an enzyme and thus bypasses the

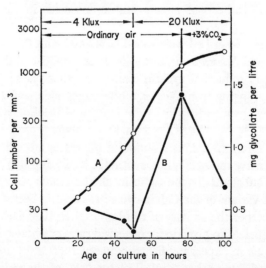

Figure 31 Growth of *Chlorella pyrenoidosa* (A) and concentration of glycollic acid in the medium (B) in relation to time and changes in carbon dioxide concentration and light intensity. With constant light and carbon dioxide concentration the con-centration of glycollic acid would have risen slowly and steadily during the growth of the culture. (Redrawn from W. D. Watt and G. E. Fogg (1966), *J. exp. Bot.*, **17**, 117.)

pool of free PGA. In either of these cases no profound modification of the Calvin cycle would be called for.

It should be noted that most of the experimental data on which Calvin's cycle is based have been obtained with algal cells at abnormally high light intensities and carbon dioxide concentrations and suspended in solutions that would not permit growth. Probably the pattern of reactions found after short periods of photosynthesis under these conditions is basically similar to that in growing cells under more normal conditions but proof of this is not as complete as one would wish it to be. Several investigators have found that at lower carbon dioxide concentrations 2-carbon compounds, particularly glycollic acid and glycine, are prominent among the substances labelled after short periods of photosynthesis with $^{14}CO_2$. The place of glycollic acid, $CH_2OH.COOH$, in the cycle is not altogether clear. It is liberated in quantity from *Chlorella* cells if these are transferred from conditions of high carbon dioxide concentration and low light intensity to low carbon dioxide concentration and high light intensity (Fig. 31), the amount being larger the greater the capacity of the cells for photosynthesis. This is in agreement with a suggestion that it is produced by breakdown of RuDP, since the amount of this might be expected to build up in the presence of high concentrations of carbon dioxide and then be in excess when this is withdrawn. Glycollic acid may be used in synthesis of sugars and amino-acids in photosynthesizing tissues.

The results just described were mostly obtained with the unicellular green algae, *Chlorella* and *Scenedesmus*, but it has been established that the mechanism of the carbon reduction cycle is similar in other plants. Twenty-seven different plants, ranging from *Nostoc*, a blue-green alga, and *Porphyridium*, a red alga, to ferns, yew and barley, were compared in photosynthesis experiments using $^{14}CO_2$ by Norris, Norris and Calvin (1955). A remarkable uniformity was found in the distribution of the tracer among alcohol-soluble compounds after five minutes' photosynthesis, indicating that the mechanism of fixation was substantially the same in all the plants tested.

An important variant of the Calvin cycle has recently been found following doubts as to whether the activity of ribulose diphosphate carboxylase in certain plants is sufficient to provide for observed rates of photosynthesis. Hatch, Slack and Johnson in papers published between 1966 and 1968 reported that in sugar cane and some other tropical plants 93% of the radioactive carbon fixed after one second of photosynthesis with $^{14}CO_2$ was in malate, aspartate and oxaloacetate, the tracer only appearing in substantial amounts in PGA after longer periods. It appears that carbon dioxide is first incorporated into these dicarboxylic acids, mainly through the agency of phosphopyruvate

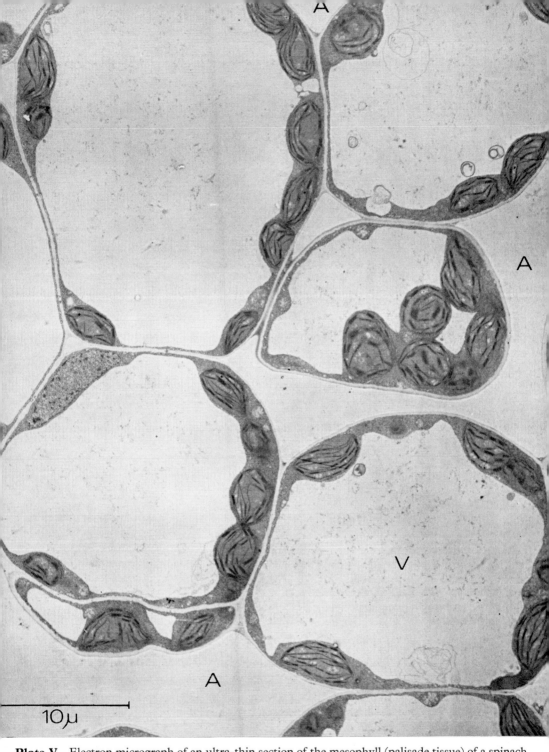

Plate V Electron micrograph of an ultra-thin section of the mesophyll (palisade tissue) of a spinach (*Spinacia oleracae*) leaf, showing the extensive intercellular air-space system (A) and the arrangement of the chloroplasts. V, the sap-filled cell vacuole. × 3800.

Plates V to XI by courtesy of A. D. Greenwood, Department of Botany, Imperial College, London.

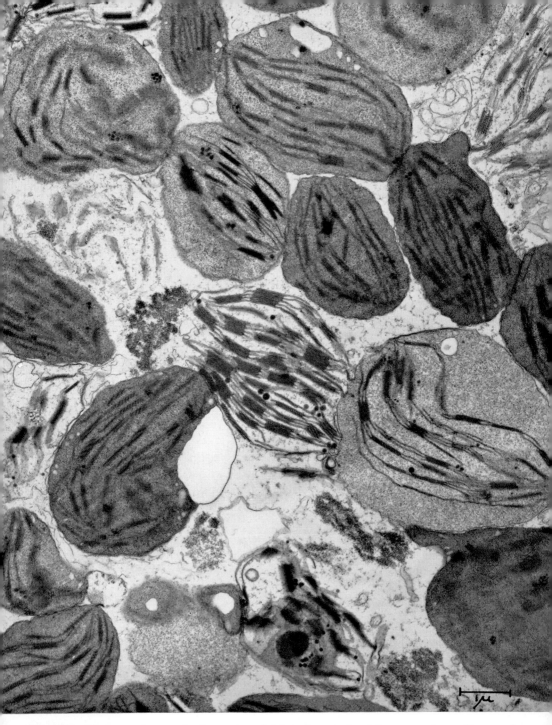

Plate VI Electron micrograph of a section of chloroplasts isolated from pea plants, *Pisum sativum*. The darker bodies are chloroplasts with envelope membranes intact, retaining the stroma. Such chloroplasts are capable of producing carbohydrate from carbon dioxide by photosynthesis. The preparation also contains chloroplast lamellar systems and cell debris. × 14,000.

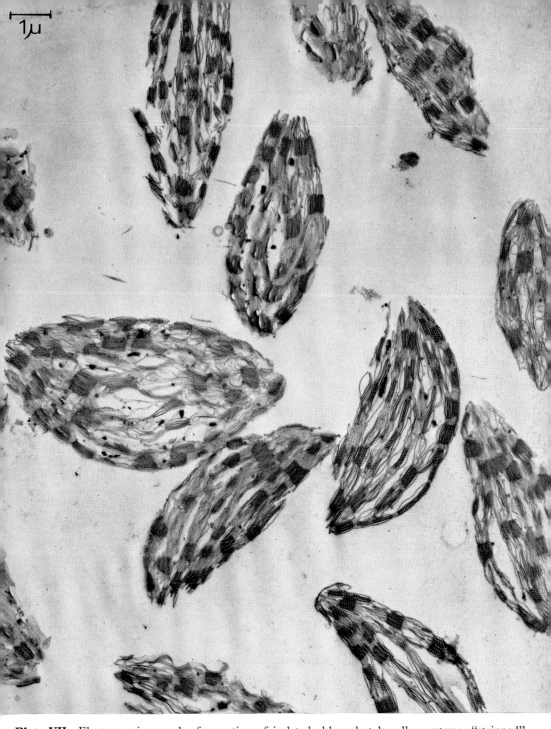

Plate VII Electron micrograph of a section of isolated chloroplast lamellar systems, "stripped" chloroplasts. A preparation similar to that illustrated in Plate VI was exposed briefly to hypotonic solution to break the chloroplast membranes and release the stroma, then purified by washing and centrifugation to remove fine debris. Such a preparation is capable of carrying out the Hill reaction and photosynthetic phosphorylation but not complete photosynthesis. ×11,000,

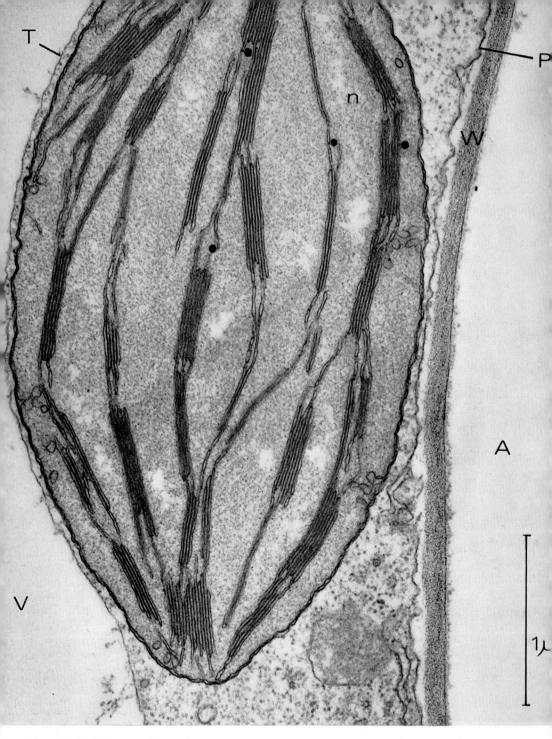

Plate VIII Electron micrograph of an ultra-thin section of a chloroplast in a cell from the mesophyll palisade tissue of spinach (*Spinacia oleracea*), glutaraldehyde fixed. The double nature of the plastid membrane and the arrangement of the lamellae in grana is shown. A, air-space of leaf; T, tonoplast (vacuole membrane); V, vacuole; W, cell wall; P, plasma membrane; n, DNA-containing "nucleoid" area of the plastid. The four small dense bodies are lipid globules. ×45,000.

carboxylase, then transferred by transcarboxylation to RuDP, thus avoiding direct carboxylation of RuDP. Thereafter the path of carbon follows the normal Calvin cycle. It is interesting that plants possessing this alternative carboxylation mechanism are amongst the most efficient in photosynthesis and growth, are not saturated even in full sunlight, and have higher temperature optima than other plants. The occurrence of this mechanism is correlated with an anatomical feature (radial arrangement of photosynthetic cells around the vascular bundles) and with a particular type of chloroplast structure.

The occurrence of the pentose phosphate carbon reduction cycle in photosynthetic bacteria

It is of particular interest that the photosynthetic bacteria, which as we have seen, have photochemical mechanisms which differ in important respects from those found in plants, nevertheless seem to have essentially the same mechanism for carbon reduction. Chromatographic analysis of the products from short periods of photosynthesis by *Rhodospirillum rubrum*, with $^{14}CO_2$ and molecular hydrogen as the hydrogen donor, showed a distribution of the tracer qualitatively similar to that in the algae. PGA was prominent among the labelled products, sugar phosphates less so, but this is understandable since carbohydrates are not stored, or utilized if supplied in the medium, by purple bacteria. It is of interest that the labelling pattern obtained when the bacterium assimilated $^{14}CO_2$ chemosynthetically in the dark using molecular hydrogen, was qualitatively rather similar to that in the light. This tallies with the finding that in *Thiobacillus* spp., which are not photosynthetic but which utilize the oxidation of various inorganic sulphur compounds as a source of energy for growth, carbon dioxide is fixed in PGA by a mechanism quite similar to that operating in green plants. Once again we see that carbon dioxide assimilation has no absolute connexion with the photochemical mechanism. In *Thiobacillus* and in *Rhodospirillum* assimilating carbon dioxide chemosynthetically the ATP and hydrogen donors required to drive the cycle are supplied by respiratory processes. When photosynthetic bacteria use organic compounds as hydrogen donors the situation becomes more complicated, since these may be assimilated directly by mechanisms dependent on the photochemical reactions.

Another carbon reduction cycle

Recently, another type of carbon reduction cycle has been found by Arnon and his collaborators in the photosynthetic bacterium *Chlorobium thiosulfato-*

philum. This operates by reversal of the Krebs tricarboxylic acid cycle. This cycle, which occurs in all aerobic organisms, plays an essential part in respiration by mediating the breakdown of pyruvic acid to carbon dioxide and water. Two reactions, which in the respiratory cycle are irreversible, are driven in the reverse direction in *C. thiosulfatophilum* by reduced ferredoxin

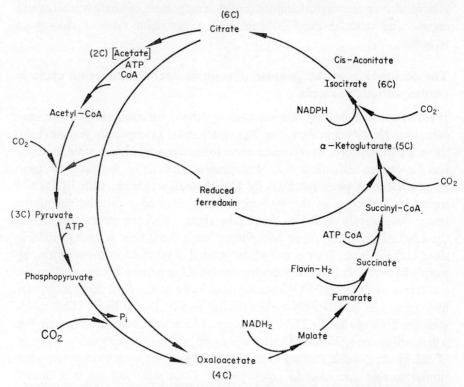

Figure 32 The reductive carboxylic acid cycle of the photosynthetic bacterium *Chlorobium thiosulfatophilum*. One turn of the complete cycle (outer sequence of reactions) results in the incorporation of four molecules of carbon dioxide. One turn of the short cycle (missing the left-hand sequence of reactions) results in the incorporation of two molecules of carbon dioxide. (Modified after M. C. W. Evans, B. B. Buchanan and D. I. Arnon, (1966) *Proc. nat. Acad. Sci.*, **55**, 928.)

and result in fixation of carbon dioxide:

26) Acetyl coenzyme A + CO_2 + ferredoxin$_{red}$ →
          pyruvate + coenzyme A + ferredoxin$_{ox}$

27) Succinyl–coenzyme A + CO_2 + ferredoxin$_{red}$ →
          α-ketoglutarate + coenzyme A + ferredoxin$_{ox}$

The reduced ferredoxin is generated photosynthetically. It will be noticed that the carboxylation is carried out directly using ferredoxin itself without the intervention of NADP as a hydrogen carrier. The place of these two reactions in the complete cycle is shown in Fig. 32. The other two carboxylations shown are ones which are reversible in the ordinary Krebs cycle, one requiring $NADPH_2$ and the other ATP. The reductive steps in the cycle are presumably also dependent on photochemically generated hydrogen donors. One turn of the complete cycle results in the fixation of four molecules of carbon dioxide to give oxaloacetic acid. A short-circuiting of the cycle may occur, as indicated in the figure, resulting in the incorporation of two instead of four molecules of carbon dioxide.

This cycle apparently occurs side by side with the pentose phosphate cycle in *C. thiosulfatophilum* but the relative importance of the two is not known. Ferredoxin-dependent carbon dioxide fixation has so far only been demonstrated in anaerobic bacteria and is not known to occur in green plants.

The location of the carbon reduction system in the chloroplast

Fractionation of isolated chloroplasts shows that the enzymes of the carbon reduction system are located in the colourless stroma (see p. 11). So far electron microscopy has not revealed any structures in the stroma which can be related to this system (Plates IX and X). Since the carbon reduction cycle has been shown to operate in solution in the test-tube (p. 8) it seems probable that the components of this multienzyme system do not need to be in definite spatial relationship to each other.

Further Reading

Bassham, J. A. (1964) Kinetic studies of the photosynthetic carbon reduction cycle. *Ann. Rev. Plant Physiol.*, **15**, 101–20.

Bassham, J. A., and Calvin, M. (1957) *The Path of Carbon in Photosynthesis.* Prentice-Hall, Englewood Cliffs, N.J.

Calvin, M., and Bassham, J. A. (1962) *The Photosynthesis of Carbon Compounds.* Benjamin, New York.

Evans, M. C. W., Buchanan, B. B., and Arnon, D. I. (1966) A new ferredoxin-dependent carbon reduction cycle in a photosynthetic bacterium. *Proc. nat. Acad. Sci.*, **55**, 928–34.

Hatch, M. D., and Slack, C. R. (1970) Photosynthetic CO_2-fixation pathways. *Ann. Rev. Plant Physiol.*, **21**, 141–62.

Stiller, M. (1962) The path of carbon in photosynthesis. *Ann. Rev. Plant Physiol.*, **13**, 151–70.

6 The flexibility of photosynthesis and its interrelations with other processes

THE picture which we now have shows the photosynthetic mechanism as consisting of a number of reaction systems more or less loosely linked together. Many of the intermediates in these reactions are also involved in non-photosynthetic metabolism. This being so, it would be expected that photosynthesis should be a rather variable process with its pattern and intensity subject to modifications according to the nature of the substrates available to the cell and its general metabolic state. This expectation is fully realized. We have already seen (p. 7) how the photosynthetic bacteria utilize various inorganic or organic hydrogen donors instead of water, and (p. 7) how certain algae which normally use water become capable of this bacterial type of photosynthesis after adaptation under anaerobic conditions. Just as different hydrogen donors may thus be substituted for water so may other substrates be substituted for carbon dioxide. The high-energy phosphate and hydrogen donors produced by the photochemical reaction are generally used in the carbon reduction cycle in the way described in the previous chapter. However, ATP and NADPH$_2$ are universal biochemical currency and may be used for a variety of other purposes. Furthermore the product of the photosynthetic reduction of carbon dioxide is not immutable and may be not only carbohydrate, which classically has been regarded as the specific product, but also any of a variety of other organic substances according to the physiological state of the cell.

Cyclic photophosphorylation in the intact plant

Although cyclic photophosphorylation is readily demonstrated in isolated chloroplasts it is not easy to prove that it occurs in living cells. However, several processes in intact plants are now known which seem to depend on cyclic photophosphorylation. These have in common two characteristics, first that they are not affected by concentrations of CMU or DCMU sufficient to inhibit photosynthetic carbon dioxide fixation, for it will be recalled that these inhibitors appear to be specific for reaction II, which produces oxygen, leaving reaction I, on which cyclic photophosphorylation depends, unaffected. Secondly, they are saturated at a much lower light intensity than that required to saturate carbon dioxide fixation (Fig. 33). These processes appear to represent the photoproduction of ATP and its utilization for processes other than carbon dioxide reduction, without concomitant production of hydrogen donor.

Under certain circumstances algae, but not usually higher plants, accumulate polyphosphate, in which phosphate groups (PO_3^-) are condensed in indefinite linear series. The so-called "volutin" granules of algae have such polyphosphates as their principal component. In *Chlorella* this accumulation is best observed in phosphate-starved cells just after the supply of phosphate has been restored. It takes place in the dark, when it is sensitive to respiratory inhibitors and requires the presence of oxygen. In the light it is more rapid and is then independent of the presence of oxygen and prevented by inhibitors which are specific for photosynthesis. In the light it is increased in the absence of carbon dioxide. It therefore appears to be the result of photophosphorylation. The linkages between the phosphate groups in polyphosphate are of the high-energy type and enzymes have been isolated from bacteria which enable the synthesis of ATP from ADP and polyphosphate. There is some evidence that in algae polyphosphate disappears when synthesis of cell material is active and it therefore appears likely to be a reserve of high-energy phosphate which accumulates when ATP production by photophosphorylation is in excess.

ATP produced by cyclic photophosphorylation may also be used to provide energy for synthesis of large molecules. The synthesis of starch in illuminated tobacco leaves appears to be an example of this. The formation of starch from glucose requires ATP and in leaf discs floated on glucose solution this can be provided equally well by aerobic respiration or photosynthetic phosphorylation, as shown by starch synthesis taking place in the light whether oxygen is present or not, but in the dark only if oxygen is present. Starch synthesis in algae may similarly be carried out at the expense of ATP produced by photophosphorylation (Fig. 33). An example of enzyme synthesis utilizing ATP produced by cyclic photophosphorylation is provided by isocitrate lyase in *Chlorella* (Fig. 33). This enzyme is only found in this alga when it has been forced to use acetate as the sole carbon source.

It is not produced when the alga grows photosynthetically on carbon dioxide, even if acetate is present, but it is formed under otherwise similar conditions if the assimilation of carbon dioxide is prevented by addition of DCMU. A further piece of evidence for cyclic photophosphorylation being involved is that isocitrate lyase synthesis is only slightly reduced when the light supplied is of wavelength 705 mμ, which favours reaction I and not reaction II and consequently permits scarcely any carbon dioxide fixation.

Arnon has pointed out that the possibility of using ATP produced by cyclic photophosphorylation for synthesis of polysaccharides, proteins and other

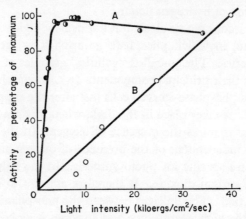

Figure 33 The effect of light intensity on the formation of the enzyme isocitrate lyase (◑, curve A), conversion of glucose to polysaccharide (●, curve A), and carbon dioxide fixation (○, curve B) under anaerobic conditions. The inhibitor DCMU was added when enzyme formation or polysaccharide synthesis was measured. (Redrawn from P. J. Syrett (1966), *J. exp. Bot.*, **17**, 641.)

such compounds means that the light energy absorbed by a plant with shut stomata, and therefore no access to atmospheric carbon dioxide, may not be entirely wasted.

The accumulation of ions by plant cells from dilute solutions takes place against the concentration gradient and therefore requires energy, which is presumably derived from ATP. Ion uptake has frequently been found to be accelerated in photosynthetic cells when they are illuminated and it is probable that this results from the coupling of ion-uptake to photophosphorylation. The production of calcium carbonate scales or coccoliths, by the flagellate *Coccolithus huxleyi*, which was discussed on page 16, is an example in which concentration of inorganic ions has been definitely related to cyclic photophosphorylation.

Photoproduction of hydrogen

We have just seen how, when production exceeds consumption, high-energy phosphate formed by cyclic photophosphorylation may be accumulated. In an analogous way, if the hydrogen donors produced by the photochemical reaction are in excess this may manifest itself by the liberation of gaseous hydrogen. For this to happen it is necessary that the organism should possess hydrogenase, the enzyme that catalyses the reaction:

28) $$AH_2 \rightleftharpoons A + H_2$$

the AH_2 in this case being photochemically produced hydrogen donor. Photosynthetic bacteria and anaerobically adapted algae possess hydrogenase and show the photoproduction of hydrogen if they are exposed to light in the presence of an appropriate hydrogen donor, such as an organic acid or an electron donor, such as thiosulphate, but in the absence of a nitrogen source which would enable them to grow. Photoproduction of hydrogen by algae reaches a maximum at a low light intensity and ceases if the intensity is increased because the hydrogenase becomes inactivated by oxidation by photosynthetically produced oxygen or an oxidizing intermediate. Higher plants lack hydrogenase and their isolated chloroplasts will only evolve hydrogen if bacterial hydrogenase is added. The photoproduction of hydrogen is accompanied by the formation of ATP.

Photosynthesis and the assimilation of nitrogen

Nitrogen is principally available to green plants as nitrate, but before it can be incorporated in the organic compounds of the cell this form must be reduced to ammonia. The hydrogen donor necessary for this reduction could conceivably be supplied directly or indirectly by the photochemical reactions of photosynthesis. Nitrate can be reduced by colourless plant cells and by green cells in the dark, but nevertheless light greatly accelerates nitrate assimilation by the latter. Thus *Chlorella*, if supplied with nitrate in the light, reduces this more rapidly than it does in the dark and evolves more oxygen than it does under otherwise similar conditions in the absence of nitrate. Warburg and Negelein (1920), who first reported this, explained their observations by supposing that nitrate is reduced in the dark at the expense of carbohydrate:

29) $$HNO_3 + 2(CH_2O) \rightarrow NH_3 + 2CO_2 + H_2O$$

and that the same process takes place in the light but then the carbon dioxide produced is used for the photosynthesis of more carbohydrate with the consequent evolution of extra oxygen. The increased rate of nitrate reduction in the light they attributed to an increased permeability of the cells in the light. However, other workers found that illuminated cells may produce oxygen in the absence of carbon dioxide if nitrate is present. Moreover, at saturating light intensities, when one must suppose that synthesis of carbohydrate is proceeding at maximum rate, supply of nitrate still increases oxygen production although uptake of carbon dioxide remains unchanged. These circumstances point to a more direct involvement of photochemical reactions than envisaged by Warburg and Negelein, so that the overall equation for the process might be put as:

30) $$HNO_3 + H_2O \xrightarrow{\text{light}} NH_3 + 2O_2$$

Further investigation of the relations of nitrate reduction and photosynthesis has not proved straightforward, partly because nitrate reduction occurs in several stages, both nitrite and hydroxylamine being produced as intermediates before ammonia appears.

31) $$HNO_3 \rightarrow HNO_2 \rightarrow NH_2OH \rightarrow NH_3$$

Each step is catalysed by a separate enzyme but there are indications that these may be associated in a single unit in some organisms. Nitrate itself does not act as a Hill reagent with spinach chloroplasts; that is to say, it seems not to be reduced directly by the photochemical reactions, but it is reduced indirectly by isolated chloroplasts in the light if NADP and nitrate reductase are added as a hydrogen-carrying system. Chloroplasts have been shown to contain the enzymes necessary for the next two steps, nitrite and hydroxylamine reductases, and isolated chloroplast grana will reduce nitrite to ammonia in the presence of added ferredoxin as a hydrogen carrier. It seems therefore that nitrite reduction can be carried out at the expense of photochemically produced hydrogen donors in isolated chloroplasts. Observations on intact algal cells are largely in agreement with this idea. In the absence of carbon dioxide the green alga *Ankistrodesmus* evolves oxygen slowly in the light if provided with nitrate, but if nitrite is supplied it evolves oxygen almost as rapidly as it does in the presence of the Hill reagent quinone. Here nitrite is evidently acting as a Hill reagent which is in agreement with the finding that its reduction takes place almost as fast at 4°C as at 15°C. In the blue-green alga *Anabaena cylindrica* the reduction of hydroxylamine, as well as that of nitrate and nitrite, is stimulated by light. 2,4-dinitrophenol, an inhibitor of oxidative phosphorylation, inhibits nitrate and nitrite reduction by algae. This has been thought to indicate an involvement of photosynthetic phosphorylation, but there is no evident need for ATP in the reduction of nitrate to ammonia itself and the effect might be due to competition between the dinitrophenol, which can act as a hydrogen acceptor, and nitrate or nitrite for the available hydrogen donor. ATP may well be necessary for the *uptake* of nitrate or nitrite into the cell (see p. 74) but this is a different matter.

The evidence is thus not altogether satisfactory, but as far as it goes it suggests that photochemically produced hydrogen donors are utilized fairly directly for nitrate reduction in most plants.

Photosynthetic bacteria and many blue-green algae are able to use the free nitrogen, N_2, of the atmosphere as a source of nitrogen. This process requires hydrogen donors and ATP and there are indications that these may be supplied photochemically. Using the heavy nitrogen isotope ^{15}N as a tracer it is possible to measure nitrogen fixation over brief intervals of time, and by this means it has been shown that the process in intact cells of the photosynthetic bacterium *Rhodospirillum rubrum* is light dependent, ceasing abruptly when the light is turned off and beginning promptly when it is

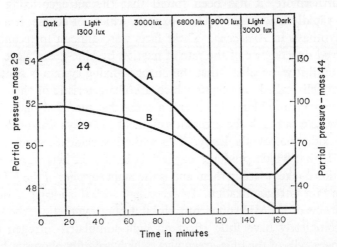

Figure 34 The effect of light intensity on the uptake of carbon dioxide (mass 44, curve A) and isotopically labelled nitrogen ($^{14}N^{15}N$, mass 29, curve B) by the photosynthetic bacterium *Rhodospirillum rubrum*. Note that the uptake of nitrogen was zero in the dark and that it increased with increasing light intensity. (Redrawn from D. C. Pratt and A. W. Frenkel (1959), *Plant Physiol.*, **34,** 333.)

switched on again (Fig. 34). In cell-free extracts of this bacterium, however, rates of nitrogen fixation are reported to be the same in the light as in the dark. Blue-green algae can fix nitrogen in the dark, but the process is stimulated in the light and its rate bears much the same relationship to light intensity as does the photosynthetic assimilation of carbon dioxide. In the absence of carbon dioxide blue-green algae are able to produce oxygen in the light if nitrogen is present. The ratio of oxygen evolved to nitrogen taken up is about 1·5, in accordance with the equation

32) $$N_2 + 3H_2O \xrightarrow{\text{light}} 2NH_3 + 1\tfrac{1}{2}O_2$$

It seems therefore that nitrogen, like nitrite, may be able to act as an alternative acceptor to carbon dioxide for photochemically generated hydrogen donors.

But once again the situation is far from simple. Pyruvic acid, $CH_3.CO.COOH$, which is a key substance for nitrogen fixation, both as a hydrogen donor and as a source of energy, in the non-photosynthetic bacterium *Clostridium pasteurianum*, has been found to be similarly important for fixation by the blue-green alga *Anabaena cylindrica*. Pyruvic acid is amongst the early products of photosynthesis (Fig. 35), but nitrogen fixation by *Anabaena* is more strongly inhibited by inhibitors specific for respiration than by those specific for photosynthesis and, also, can proceed slowly in the dark. Furthermore, it has been found that the nitrogen-fixing enzyme system is rapidly inactivated by free oxygen, which, of course, is a product of photosynthesis in *Anabaena*. These facts were difficult to reconcile with the observed dependence of the rate of fixation by blue-green algae on light intensity until it was realized that the nitrogen-fixing system is contained in the special cells called *heterocysts* which are characteristic of these species. Heterocysts contain chlorophyll *a* and a somewhat modified photosynthetic lamellar system but lack the accessory pigment phycocyanin and cannot fix carbon dioxide photosynthetically. This and other evidence shows that light reaction II, that which produces oxygen, is lacking in heterocysts but that light reaction I takes place in them and is the main supplier of the ATP which is necessary for nitrogen fixation. In agreement with this idea it is found that concentrations of DCMU sufficient to inhibit photosynthetic carbon dioxide fixation completely have little effect on light-stimulated nitrogen fixation. The ordinary cells of the blue-green alga, which are fully photosynthetic and evidently incapable of nitrogen fixation under aerobic conditions, supply the heterocysts with hydrogen donor and carbon skeletons, perhaps in the form of pyruvate, and in return receive fixed nitrogen from them. The blue-green algae thus seem to have solved the problem of combining nitrogen fixation with photosynthesis by reverting, in these special cells, to the bacterial type of photosynthesis.

The photoassimilation of organic substances

In normal photosynthesis the starting material is the most highly oxidized form of carbon, the dioxide, but there seems to be no compelling reason why this should invariably be the carbon source. Various organic substances in which the carbon is more or less reduced might equally readily be built up into cell constituents at the expense of photochemically generated assimilatory power. It has been known for a long time that the photosynthetic growth of most algae can be stimulated by supply of a suitable organic substrate such as acetate or glucose. A feature of this stimulation is that it is evident only under conditions of light limitation and that the stimulation is usually only such as to bring the growth rate up to the maximum attainable at saturating light intensity with carbon dioxide as the sole carbon source. This may be taken as an indication that the assimilation of the organic substrate is photosynthetic, since if it were not the effect would perhaps be additive at saturating light intensity. More definite evidence is provided by studies of the effect of manganese on growth. In the dark the growth of *Chlorella* on glucose is the same in the absence as in the presence of a trace of this element. However, manganese is essential for photosynthesis (p. 57) and growth in the light is much reduced in its absence irrespective of whether the carbon source is carbon dioxide or glucose. This is strong evidence that assimilation of glucose in the light is a photosynthetic process.

This possibility may be further investigated by supplying an alga with a ^{14}C-labelled organic substrate and comparing the distribution of the radiocarbon after short periods in light and dark, using methods similar to those for study of the early products of photosynthesis. With various algae and different substrates it has been shown that the pattern of labelling is that characteristic of the photosynthetic carbon reduction cycle rather than that of respiration. Glucose, for example, enters directly into the carbon reduction cycle in *Chlorella* in the light, giving rise to Calvin cycle intermediates rather than Krebs cycle intermediates. Thus the organic substrate is not broken down via the Krebs cycle to carbon dioxide which is then assimilated photosynthetically.

The biological advantage of photoassimilation of organic substrate may be considerable in dimly lit situations. By starting with a reduced source of carbon the limited assimilatory power from the photochemical reactions may be used to provide for a much greater amount of growth than would otherwise be possible. It is not altogether surprising therefore that some photosynthetic organisms are adapted to this type of metabolism.

The photosynthetic bacterium *Chromatium* can grow in the absence of carbon dioxide in the light if it is provided with acetate. Acetic acid, CH_3COOH or $(CH_2O)_2$, is already at the reduction level of carbohydrate and therefore does not require a major source of reducing power for its assimilation. In fact it seems that the main rôle of light in acetate assimilation by

Chromatium is to provide ATP by cyclic photophosphorylation, since it can be replaced by supplied ATP and cell-free bacterial extracts thereby enabled to assimilate acetate in the dark. A similar type of metabolism is met with in a green alga, *Chlamydomonas mundana*, found in sewage lagoons, which, of course, are rich in organic matter including acetic and other fatty acids. This alga can grow photosynthetically with carbon dioxide, but in the presence of acetate it grows vigorously in the light without, however, producing oxygen or fixing carbon dioxide to any great extent. This photoassimilation of acetate again seems largely dependent on photophosphorylation, since it is not greatly affected by the inhibitor DCMU at concentrations sufficient completely to inhibit oxygen production, and therefore the production of hydrogen donors, in normal photosynthesis. Like those of *Chromatium*, cell-free extracts of this alga assimilate acetate if ATP is added. The alga is not able to grow on acetate in the dark, however, because of its limited capacity for oxidative phosphorylation. Another green alga, *Chlamydobotrys* sp., is quite unable to grow by normal photosynthesis with carbon dioxide and is completely dependent on photoassimilation of acetate. In *Chromatium* adapted to acetate as a carbon source the production of ribulose diphosphate carboxylase (see p. 64) is suppressed. Likewise in *Chlamydomonas mundana* acetate strongly inhibits the formation of several enzymes of the carbon reduction cycle although it does not affect the chlorophyll content or photochemical activities of the cells. Control is thus exerted at the level of enzyme production, and presumably the inability of *Chlamydobotrys* to assimilate carbon dioxide photosynthetically is due to the irreversible loss of the capacity to produce the necessary enzymes.

The products of photosynthesis

For a long time it was believed that carbohydrate is the product of photosynthesis in the sense that the carbon fixed necessarily appears in this form before being incorporated in fats, proteins and other classes of substance. Starch frequently appears quickly on illumination of starch-free leaves and in mature plants gains in weight occurring as a result of photosynthesis can be largely accounted for in terms of carbohydrate. A great number of measurements on all types of plant of the photosynthetic quotient ($Q_P = \Delta O_2 / -\Delta CO_2$) also lend support to this view, since they lie close to unity as is to be expected if carbohydrate is the product. It is saddening that for nearly fifty years plant physiologists did not realize how biased this evidence is. It is most convenient in experiments on photosynthesis to use high light intensities and as material either mature leaves or algae. If the photosynthesis of algae is being studied growth is an unwanted complication, so that the cells are suspended in a medium, lacking a source of nitrogen, in which they cannot multiply. In both cases conditions are such that carbohydrate synthesis is favoured and protein synthesis is prevented. If young growing tissues or algal cells suspended in complete growth medium are used then values of Q_P of

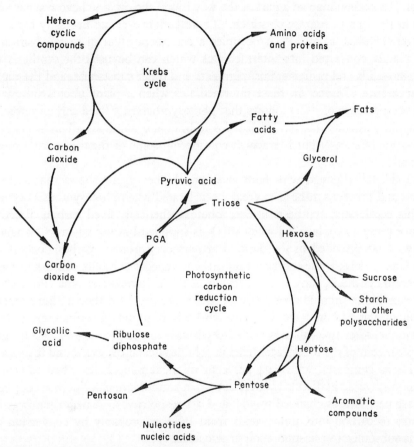

Figure 35 A general scheme showing some of the interrelations between the photosynthetic carbon reduction cycle, the Krebs tricarboxylic acid cycle, and the synthesis of various types of compound. (Modified after O. Holm-Hansen *et al.* (1959), *J. exp. Bot.*, **10**, 109.)

about 1·1, corresponding to protein synthesis, are observed if ammonium salts are the source of nitrogen. Similarly if algal cells are brought into the condition in which fat accumulation occurs then values of Q_P, sometimes as high as 3·3, appropriate to this will be obtained with them.

Now that we know more of the detailed mechanism of the photosynthetic fixation of carbon it is clear that there is no substance which can be described as the product of photosynthesis in any special sense. The first stable product of the fixation, in which, however, the carbon has not yet undergone reduction, is phosphoglyceric acid, PGA. To maintain the reduction cycle five PGA molecules out of every six must be returned into it but the sixth need not. Its carbon may go a part of the way round the cycle to hexose sugar and then perhaps to sucrose or starch. There is, however, no necessity for it to enter the cycle at all. PGA occupies a central position in metabolism and is readily converted into intermediates which can be used for synthesis of amino-acids and proteins, fats, pigments and other substances, and the rapid appearance of tracer in these intermediates when a plant photosynthesizes in the presence of $^{14}CO_2$ shows that photosynthetically fixed carbon finds its way into them directly. These possibilities are summarized in a general diagram (Fig. 35), but it is worth considering some of them in a little more detail.

Evidence obtained with both algae and higher plants shows that amino-acid and protein synthesis is greatly accelerated when photcsynthesis occurs. This accelerated synthesis utilizes photosynthetically fixed carbon directly, since proteins are labelled when $^{14}CO_2$ is supplied but not when ^{14}C-labelled sugars are provided in the dark. Furthermore, protein synthesis occurs in isolated chloroplasts, and ribosomes, the organelles specially concerned with protein synthesis, have been detected in the chloroplast stroma (Plate IX). If conditions suitable for protein synthesis are provided then a high proportion of the carbon fixed in photosynthesis is found in amino-acids. For example, after two minutes of photosynthesis in the presence of $^{14}CO_2$ only 9·9 per cent of the total tracer fixed in soluble substances was found in amino-acids as compared with 64·4 per cent in sugar phosphates when *Chlorella* was suspended in distilled water, but when ammonium chloride was supplied these percentages changed to 56·8 and 7·1 respectively. A major pathway of entry of carbon into amino-acids from PGA is probably by conversion of this substance to phospho-enolpyruvic acid:

33) $CH_2O-P.CHOH.COOH \rightarrow CH_2:CO-P.COOH + H_2O$

An amino group may be added to this giving the amino-acid alanine

34) $CH_2:CO-P.COOH + NH_4^+ + NADPH_2 \rightarrow$
                    $CH_3.CH(NH_2).COOH + NADP + P$

the $NADPH_2$ being presumably supplied by the photochemical reactions. Phospho-enolpyruvic acid may also be carboxylated to give oxaloacetic acid, a four-carbon acid from which another amino-acid, aspartic acid

($COOH.CH_2.CH(NH_2)COOH$), may be formed. Alanine and aspartic acid are in fact among the first intermediates to be labelled during photosynthesis with $^{14}CO_2$ (Plate II*b*). Other amino-acids may also be derived, but less directly, from PGA.

Fat synthesis may also take place directly from PGA. Accumulation of fats is characteristic of algae which have exhausted their supply of nitrogen. For example, the diatom *Navicula pelliculosa*, which may have as much as 35 per cent fat on a dry weight basis when nitrogen deficient, has been found to incorporate up to 70 per cent of the total radiocarbon supplied as $^{14}CO_2$ into fats in a two-minute period of photosynthesis. When nitrogen was supplied to similar cells as ammonium chloride the percentage fell to 7. The synthesis of fatty acids in chloroplasts isolated from sunflower plants has been demonstrated. All the possible pathways of fat synthesis start with acetyl coenzyme A, which is probably produced mainly by oxidation of pyruvic acid, and as we saw in the last paragraph phospho-enolpyruvic acid is readily derived from PGA. It is significant that acetate supplied to algal cells in the light is largely incorporated into fats. The availability of ATP from photophosphorylation enables this to be converted to acetyl phosphate and from this acetyl coenzyme A can be formed. Fatty acids are highly reduced substances and their synthesis from acetyl coenzyme A requires hydrogen, which again is probably donated by $NADPH_2$ produced in the photochemical reactions.

Glycollic acid was mentioned in the previous chapter (p. 68) as an intermediate of photosynthetic carbon fixation which may escape from *Chlorella* cells into the surrounding medium. This substance is perhaps sometimes the major product of photosynthesis, since plankton algae have been found to liberate as much as 95 per cent of the carbon fixed in photosynthesis in extracellular form. Such massive release occurs in the surface waters of the sea and of lakes on clear days when the light intensity is sufficiently high to inhibit photosynthesis. In fact the proportion of the carbon fixed which is liberated in extracellular form is more the greater the inhibition of photosynthesis (Fig. 36). This may be looked on as overflow mechanism. As we have seen (p. 32), the rate of photosynthesis of plankton algae is largely dictated by the environment and at high light intensity, especially perhaps in waters poor in the mineral elements needed for growth, it is conceivable that carbon is fixed at a greater rate than it can be dealt with by the synthetic mechanisms of the cell. High light intensities, especially if there is an appreciable proportion of ultra-violet, appear to have a more damaging effect on growth mechanisms than on the photosynthetic apparatus itself. In these circumstances more carbon will be fixed than can be built up into cell material or reserve products and the excess appears as extracellular products such as glycollate. Possibly glycollate may be reabsorbed and used to maintain growth at low light intensities, so that it might be looked on as an extracellular reserve product, but as yet there is no evidence for this and it is more likely that the glycollate released is utilized by bacteria.

There are indications that the product of photosynthesis may vary according to the wavelength of the light which illuminates the plant. There is still some uncertainty about the reality of these effects but we might expect

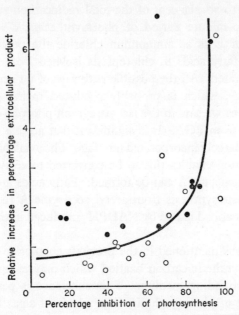

Figure 36 The effect on the proportion of photosynthetic product liberated in extracellular form of light inhibition of photosynthesis in natural waters; ○, in Windermere, ●, in Tring reservoirs. (Data of W. D. Watt (1966), *Proc. Roy. Soc.* B, **164**, 521.)

something of this sort to happen, since the two light reactions of photosynthesis have different action spectra. Illumination with different wavelengths might thus alter the balance between their products, ATP and $NADPH_2$, and this might reasonably be expected to have some influence on the relative amounts of carbon following the various biochemical pathways.

Transients

Although dark and light metabolism have many enzymes and intermediates in common, different metabolic pathways are important in each case and in most plants respiration and photosynthesis are carried out in distinct regions of the cell. Consequently when a plant is switched from dark to light there is a period of adjustment in which concentrations of intermediates alter and their diffusion gradients from one metabolic site to another are changed, before they settle down to the new steady state with all reactions in balance. Similar transitions must occur to a greater or lesser extent whenever an environmental factor such as light intensity or carbon dioxide concentration alters. One obvious effect of the imbalance of reactions is that the rate of photosynthesis characteristic of the new conditions is not attained immediately, but in addition there are various transient or induction effects, not

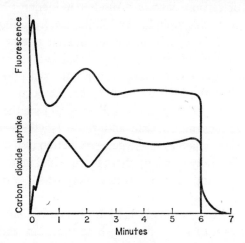

Figure 37 Induction phenomena in photosynthesis: time course of fluorescent yield and carbon dioxide uptake in the first few minutes after the beginning of photosynthesis by wheat leaves in 0·24% carbon dioxide and strong light after 10 minutes' dark rest. (Redrawn from E. D. McAlister and J. Myers (1940), *Smithson. Misc. Coll.*, **99** (6).)

characteristic of steady state photosynthesis. It may be imagined that these transients are particularly sensitive to variations in environment and physiological state of the cells, and it is perhaps for this reason that this aspect of photosynthesis is one of the most confusing. However, some of the more striking of these effects should be mentioned as their study has contributed substantially to our knowledge of photosynthesis.

All plants investigated show a low rate of carbon dioxide uptake immediately following change from dark to light and do not attain a steady rate until they have been illuminated for several minutes. The rise to the steady rate may be irregular (see Fig. 37) and with some plants, *Chlorella* for example, there may be actual liberation of carbon dioxide, a carbon dioxide "gush", in the first few seconds after the light is turned on. The explanations of these effects are not certain but, as we have already noted (p. 64), the carbon dioxide acceptor, RuDP, is used up by carboxylation to give PGA after light has been switched off, with the result that its concentration needs to be built up again on return to light and there will be a corresponding period of reduced uptake of carbon dioxide. An effect which may be superimposed on this depends on a peak in ATP concentration, reached 15–30 seconds after the beginning of illumination, resulting from photophosphorylation temporarily producing ATP in excess of demand. Since synthesis of ATP consumes inorganic phosphate ions the protoplasm will become more alkaline and this will cause increased uptake of carbon dioxide. Another induction phenomenon is increased fluorescence of chlorophyll. As will be seen from Fig. 37, the fluorescence curve tends to mirror that for carbon dioxide uptake. This illustrates a point made in an earlier chapter (see p. 44) that fluorescence of chlorophyll increases when the light energy absorbed is not used photochemically. In this case it seems that the high initial fluorescence occurs when the carbon reduction cycle is not yet running at full rate, so that the products of the photochemical reactions accumulate and these reactions are consequently suppressed. Transient changes in oxygen production also occur when plants are illuminated. Generally speaking there is a gradual increase in rate during the first few minutes, but oxygen "gushes" may be superimposed on this in the first few seconds.

The existence of these various induction phenomena support our concept of photosynthesis as a series of intermeshing reactions and it is significant that the Hill reaction and other photochemical reactions of isolated chloroplasts, which represent partial reactions of photosynthesis, do not show them.

Apart from short-term effects, which largely reflect changes in the photosynthetic mechanism itself, there are slower changes, which depend on redistribution of intermediates throughout the cell, involved in transfer from light to dark and *vice versa*. Clearly both long- and short-term effects may affect growth and a plant which experiences too frequent changes from light to dark may not attain the balance of metabolism most favourable for growth. Most plants grow best if given a light/dark cycle similar to the natural alternation of day and night. At the other extreme good growth can be achieved with alternating flashes and dark periods of a fraction of a second up to a few seconds in length. With extremely brief flashes the increase in photosynthetic efficiency discussed on page 5 comes into play. In between, with alternating dark and light periods of a minute or so in length, there is a strong inhibitory effect on growth. An extreme example is afforded by the flowering plant *Cosmos sulphureus*, growth of which is completely suppressed by this treatment. Evidently illumination for one minute is just sufficient to bring the photosynthetic mechanism into full gear and one minute's dark is sufficient to set it back to the starting point again, with the result that no useful production of food materials is achieved. A less extreme example of similar effects of different rates of intermittent illumination is afforded by results obtained with diatoms, growth of which is about half of that obtained with twelve hours' light, twelve hours' dark, when one minute light, one minute dark periods are given (Fig. 38).

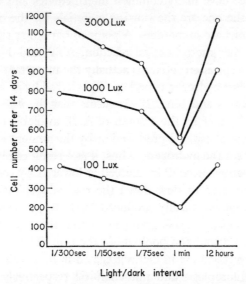

Figure 38 Effect of intermittent light, ranging from alternating light and dark periods of 1/300th of a second each to 12 hours' light and 12 hours' dark, on the growth of a freshwater diatom in 14-day-old cultures exposed to different light intensities. (After H. Behrend (1948), *Arch. Mikrobiol.*, **14**, 531.)

Photosynthesis and respiration

The interrelations of photosynthesis and respiration deserve special consideration because of their importance in the interpretation of experimental results. Everyone who sets out to measure the rate of photosynthesis has to decide what correction to make for respiration. The rate as measured, the *nett* rate, is presumably less than the actual rate because the reverse process, respiration, has been going on at the same time. Usually the correction is made on the assumption that respiration proceeds at the same rate in photo-synthesizing cells as in the same cells in the dark. Thus the amount of oxygen consumed under otherwise similar conditions but in the dark is added on to the amount of oxygen produced by the same or similar cells during an equal period in the light, to give the *gross* rate of oxygen production in photosynthesis.

This procedure has the merits of simplicity and of yielding more consistent results than any other, but until quite recently its validity could not be tested in any unequivocal way. From much that has already been said it will be realized that it is somewhat unlikely that photosynthesis and respiration can be as completely independent of each other as is assumed in using this method. The two processes have many common intermediates and enzymes and the end-products of the one are the starting materials for the other, so that the possible interactions are numerous. A point of particular significance is that the products of the photochemical reaction, ATP and $NADPH_2$, are key intermediates in respiration (NAD is actually the important hydrogen acceptor in respiration but hydrogen is perhaps transferred by an enzyme-catalysed reaction between this and NADP). It seems that the rate of respiration is usually controlled either by the amount of ADP available to act as acceptor for the high-energy phosphate produced or by the amount of NAD available to act as acceptor for the hydrogen. Thus, if as a result of photophosphorylation the concentration of ADP in the cell is lowered by conversion to ATP this would be expected to slow down the rate of respiration. Transfer of hydrogen from photochemically produced $NADPH_2$ to NAD would have a similar effect. A fallacy in this argument is that these intermediates do not necessarily exist in a common pool within the cell. The processes of photosynthesis and respiration are in fact segregated in most cells in quite distinct organelles, the chloroplasts and mitochondria respectively (Plate IX), and these may have their own independent pools of ADP and NADP or NAD. On the other hand, electron micrographs sometimes show the two kinds of organelle in intimate contact and there is evidence from experiments with isotopically labelled intermediates that exchange between chloroplasts and cytoplasm is often rapid. Finally, the photosynthetic evolution of oxygen and the respiratory uptake of oxygen are terminal processes carried out by distinct and probably rather different enzyme systems.

Such are the possible complexities that it is difficult to assess from considerations of this sort what the outcome is likely to be. Fortunately, with isotopic tracers it is now possible to make direct observations. One such approach has been to use isotopic labelling to distinguish between oxygen taken up and given out. Some results obtained in this way with a tobacco (*Nicotiana tabacum*) leaf are shown in Fig. 39. The leaf was supplied with air enriched with the heavy oxygen isotope, ^{18}O, and changes in the amount of oxygen molecules of different mass were followed during periods of light and dark using a mass-spectrometer. The proportion of molecules of mass 32 (ordinary oxygen $^{16}O^{16}O$) decreased in the dark as a result of uptake by respiration, but increased in the light because of liberation of oxygen by photolysis of water containing ^{16}O only. The proportion of molecules of mass 34 ($^{18}O^{16}O$), however, decreased in both light and dark, with only a slight change in rate due to the altered ratio of the two kinds of molecules after illumination. This shows that respiration continued at the same rate in light and dark. The clean separation between incoming and outgoing oxygen is rather astounding; one would expect that oxygen evolved in photosynthesis

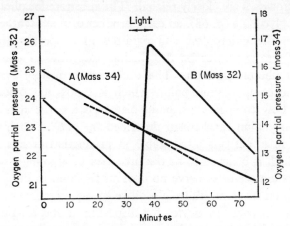

Figure 39 Respiration during photosynthesis in a leaf of tobacco (*Nicotiana tabacum*) as recorded with a mass-spectrometer. The oxygen in the atmosphere was labelled with the isotope ^{18}O, giving molecules, $^{16}O^{18}O$, of mass 34. These were taken up at the same rate in the light and dark (the change in slope of curve A was due to the altered proportion of molecules of mass 34 following illumination). Photosynthesis evolved oxygen having mass 32 from water; curve B shows uptake of these molecules in the dark and their production in the light. (Redrawn from A. H. Brown (1953), *Amer. J. Bot.*, **40**, 719.)

would be used to some extent in respiration, giving rise to an apparent reduction in respiration rate, and, in fact, results are not often as clear-cut as this. There is evidently variation according to species, the physiological state of the organism, and the conditions to which it is exposed. With the blue-green alga *Anabaena*, for example, the ^{18}O method shows a range of effects varying from complete suppression of respiration in the light, even at low intensities below the compensation point, to a 2·5-fold increase in strong light.

Such results, together with a great deal of other evidence, show that relatively weak illumination suppresses dark respiratory processes and stimulates another type of respiration, *photorespiration*. This may be detected under suitable conditions, using isotopic tracers, as increased carbon dioxide output or oxygen uptake and it is distinguished from the mitochondrial respiration characteristic of plants in the dark by the way in which it goes on increasing with oxygen concentration up to 100 per cent whereas mitochondrial respiration is saturated at around 2 per cent oxygen concentration. The results with the tobacco leaf mentioned above were obtained at low oxygen concentration so that photorespiration was not pronounced. Photorespiration increases with increasing light intensity and is inhibited by DCMU. In general, factors which affect photosynthesis affect photorespiration similarly and it is clear that the two processes are closely related. The substrate for photorespiration appears to be glycollate (p. 68), an early product of photosynthesis which is liberated from the chloroplast and oxidized in a specific organelle, the *peroxisome*. There appear to be two distinct categories of plants, showing high and low photorespiration. The difference is mainly evident in the carbon dioxide compensation point, which is 0·005 and 0·0005 per cent respectively. Low compensation point plants possess the alternative, more efficient, carboxylation mechanism described on p. 68. Their low rate of photorespiration is perhaps attributable to production of less ribulose diphosphate, which is believed to be the precursor of glycollate.

Photorespiration seems to serve no obvious function in the plant, unless, like the liberation of extracellular glycollate by aquatic algae (p. 83), it can be regarded as an overflow device for disposing of surplus photosynthetic products. It has been suggested that the productivity of crop plants might be increased if photorespiration could be selectively inhibited, but perhaps it is over-optimistic to expect that interference with such a complex system would have such a straightforward effect.

The biological functions of photosynthesis and mitochondrial respiration seem to be identical; both provide assimilatory power in the form of high-energy phosphate and hydrogen donors and a pool of intermediates which can be used as building units for more complex molecules. Photosynthesis and respiration thus have interchangeable rôles in the plant. This is evident in the fact that the numbers of the organelles concerned usually show an inverse relation. Colourless tissues have cells with numerous mitochondria, but these are infrequent in cells with abundant chloroplasts. Some algae, called obligate phototrophs, are only able to grow photosynthetically. Their inability to grow in the dark on organic substrates may arise from any of various causes, but one of these is evidently a deficiency in the respiratory mechanism which prevents the production of ATP by oxidative phosphorylation. Photophosphorylation is then the only source of ATP. One obligate phototroph, the blue-green alga *Anacystis nidulans*, has a particularly low rate of respiration. Examples of plants which have come to depend entirely on respiration and have lost the capacity to form chloroplasts scarcely require mention. The flagellate alga, *Ochromonas malhamensis*, is, however, worth noting as an intermediate type. This is capable of taking in particulate food, like an animal, as well as growing saprophytically or photosynthetically. Its powers of photosynthesis are feeble at the best—with light and carbon dioxide alone it grows at only about one-fifth of the rate which is attained if glucose is available. Myers and Graham (1956) described *Ochromonas* as "a very primitive animal which has retained only enough of its photosynthetic apparatus to sustain it between bites".

Variation of photosynthesis with stage of development

Finally in this chapter we should look briefly at the way in which variations in photosynthesis are related to the development of the plant. Firstly it may be noted that there are quite large variations in rate per unit amount of photosynthetic tissue or cells. In terms of carbon dioxide fixed per unit area under standard conditions mature leaves are more active than young expanding ones or ageing ones (Fig. 40). Cells from actively growing cultures of algae may photosynthesize at ten or twenty times the rate of cells from cultures in which the nitrogen supply has been exhausted and which have ceased growing. There are also daily variations. When subjected to alternating periods of twelve hours light and twelve hours dark but otherwise constant conditions, the green alga *Hydrodictyon* shows an increase in rate of photosynthesis to a peak some four hours after the beginning of the light period, after which the rate declines at the end of the period of illumination to about a third of the maximum. Similar variations evidently occur in the leaves of higher plants, but unless special precautions are taken to maintain all factors constant they are usually obscured by changes in leaf water content and stomatal aperture which affect the rate of photosynthesis indirectly.

Such changes in rate are probably symptomatic of changes in the balance between the various component reactions of photosynthesis, first one, then another, reaction limiting the rate and the most efficient relationship between them perhaps being achieved only momentarily. Evidence of such changes in pattern of photosynthesis during the growth of individual algal cells has come from study of synchronous cultures. These are cultures in which by appropriate treatment the cell division cycle in the population has been brought into step. Usually in an actively growing population cell divisions are out of step and a sample taken at any time will contain cells in all stages— those just produced by division, those ripe to divide and all intermediate stages—so that biochemical determinations give only average values and one cannot tell from them what sort of variations in activity go on during the division cycle. With a synchronous culture on the other hand measurements of biochemical activity or cell composition made at successive times give a picture of what goes on as an individual cell develops. In this way it has been found that in general the photosynthetic capacity of the daughter cells produced by division of a mature algal cell is low, but that it rises to a maximum in the enlarging cells then declines as they in turn become ripe to divide. At light saturation the maximum photosynthetic activity may be four or more times the minimum. With one strain of *Chlorella* the photosynthetic quotient was found to rise from about unity for the young cells to over three for cells ripe to divide. This suggests that as the cells mature the photosynthetic product changes from protein or carbohydrate to highly reduced substances such as fats. Analyses of the cells support this conclusion but with other strains similar changes have not been observed. Recently, it has been found with the green alga *Scenedesmus* that, whereas the quantum requirement for photoreduction by anaerobically adapted cells remained nearly constant around 20 during the life-cycle, that for photosynthesis rose from 10·7 in young cells to 16·6 in mature cells. Since photosynthesis requires the participation of both light reactions whereas photoreduction is supposed to require only light reaction I, this seems to show that there is a change in the balance between the two light reactions during the development of the individual cell of *Scenedesmus*.

Conclusion

We have seen that there is definite evidence that the assimilatory power generated by the photochemical reactions may be used in the normal life of plants for certain purposes quite other than the production of carbohydrate from carbon dioxide. Indeed there are indications that all the possible fates which can be envisaged for the hydrogen donors and high-energy phosphate produced by these reactions are realized in one or other kind of plant under appropriate conditions. There is no fixed relationship between the component reactions of photosynthesis and no unique biochemical pathway for the carbon fixed. The old definition of photosynthesis as the production of carbohydrates from carbon dioxide by green plants in the light is thus inadequate. In view of the flexible relations which exist between the photochemical reactions and the other life processes of plants we must recognize that the essential feature of photosynthesis is the conversion of radiant energy to potential chemical energy, and that the nature of the substance in which this potential chemical energy is stored is a quite secondary matter.

Further Reading

Betts, G. F., and Hewitt, E. J. (1966) Photosynthetic nitrite reductase and the significance of hydroxylamine in nitrite reduction in plants. *Nature*, **210**, 1327–9.

Calvin, M., and Bassham, J. A. (1962) *The Photosynthesis of Carbon Compounds.* Benjamin, New York.

Fay, P. (1970) Photostimulation of nitrogen fixation in *Anabaena cylindrica. Biochim. Biophys. Acta*, **216**, 353–6.

Fogg, G. E. (1959) Nitrogen nutrition and metabolic patterns in algae. *Symp. Soc. exp. Biol.*, **13**, 106–25.

Gaffron, H. (1960) Energy storage: Photosynthesis. In *Plant Physiology* edited by F. C. Steward, vol. IB, pp. 3–277. Academic Press, New York and London.

Jackson, W. A., and Volk, R. J. (1970) Photorespiration. *Ann. Rev. Plant Physiol.*, **21**, 385–432.

Kessler, E. (1964) Nitrate assimilation by plants. *Ann. Rev. Plant Physiol.*, **15**, 57–72.

Senger, H., and Bishop, N. I. (1967) Quantum yield of photosynthesis in synchronous *Scenedesmus* cultures. *Nature*, **214**, 140–2.

Senger, H., and Bishop, N. I. (1969) Emerson enhancement effect in synchronous *Scenedesmus* cultures. *Nature*, **221**, 975.

Simonis, W. (1960) Photosynthese and lichtabhängige Phosphorylierung. In *Encyclopedia of Plant Physiology* edited by W. Ruhland, vol. V (1), pp. 966–1013.

Stewart, W. D. P., Haystead, A., and Pearson, H. W. (1969) Nitrogenase activity in heterocysts of filamentous blue-green algae. *Nature*, **224**, 226–8.

Syrett, P. J. (1966) The kinetics of isocitrate lyase formation in *Chlorella*: Evidence for the promotion of enzyme synthesis by photophosphorylation. *J. exp. Bot.*, **17**, 641–54.

Veen, R. van der (1960) Induction phenomena. In *Encyclopedia of Plant Physiology* edited by W. Ruhland, vol. V (1), pp. 675–88.

Watt, W. D. (1966) Release of dissolved organic material from the cells of phytoplankton populations. *Proc. Roy. Soc. B*, **164**, 521–51.

Wiessner, W. (1970) Photometabolism of organic substances. In *Photobiology of Microorganisms* edited by P. Halldal, pp. 95–133. John Wiley & Sons Ltd., London and New York.

7   *Photosynthesis and life: past, present and future*

LIFE as we know it is absolutely dependent on photosynthesis. Were this process to cease it could not be adequately replaced by any other and, such is the rate at which non-photosynthetic organisms consume plant material and each other, their stock of food would rapidly be depleted and the higher forms of life at least would become extinct within twenty-five years or so. At present we have nearly steady state conditions with the production of organic matter balancing its consumption so that the concentrations of oxygen and carbon dioxide in the atmosphere are maintained at almost constant levels. Conditions cannot always have been like this and it is of interest to consider how and when photosynthesis began and what part it played in modifying the primitive environment of the Earth and thus determining the course of evolution. We must also consider what effect Man's activities may have on photosynthesis in the future and what uses he may make of it as increasing knowledge of its mechanism enables him to control and direct it with greater effect.

Photosynthesis and the origin of life

The atmosphere of the primitive Earth was different in composition from that existing today. It seems probable that the Earth at the terminus of its formation had an atmosphere containing water, hydrogen, ammonia and some hydrogen sulphide, but no oxygen or carbon dioxide. Organic substances are formed in such a mixture if it is exposed to electric discharges or ultra-violet light, and in an experiment reported in 1955 Miller showed that among these are many of biological importance such as fatty acids and amino-acids.

When ultra-violet radiation is used this synthesis is a photochemical one but the radiation provides only activation energy, the starting substances are already reduced and there is no net accumulation of potential chemical energy as in photosynthesis in plants. A gas mixture of the composition which we believe the primitive atmosphere to have had is transparent to ultra-violet, and we may suppose that as a result of photochemical reactions and concentration by evaporation or adsorption on clays an organic soup was formed in places on the surface of the primitive Earth. In this, after a long period of chemical evolution, more elaborate compounds might have been built up, some of them having the property of self-replication and by selection and combination between reaction systems something having the characteristics of life gradually evolved. We need not spend time in conjecture about details; the point that concerns us here is that according to this theory, which was advanced independently by Haldane and Oparin and which is now widely accepted, the first organisms were not photosynthetic but heterotrophic, gaining their energy by anaerobic conversion of pre-existing organic substances.

Formation of organic substances promoted by ultra-violet light would, however, decrease at length. This would be because hydrogen, being a light element, can escape from the earth's gravitational field and, at the point at which the atmosphere lost excess free hydrogen, oxygen would begin to accumulate as the result of the decomposition of water by ultra-violet light. Ozone, derived from oxygen, absorbs ultra-violet light strongly, and so the amount of this type of radiation reaching the Earth's surface would be drastically reduced. This chain of events would be self-regulating, since the reduction in penetration of ultra-violet into the atmosphere would slow down the decomposition of water, so that the concentration of oxygen would be well below 1 per cent.

In the absence of any other process producing organic material the first, heterotrophic, life would have petered out and may well have done so on numerous occasions until among the organic substances formed on the primitive earth there appeared pigments capable of initiating chemical reactions after having become excited by absorption of visible radiation. The porphyrins—to which class chlorophyll itself belongs—are such pigments which perhaps played a decisive part from the very beginning because they can be easily formed from the simpler substances, acetic acid and glycine, produced in Miller's experiment. Porphyrins are excellent photosensitizers in laboratory experiments, promoting various electron or hydrogen transfers and direct oxidations with organic substrates. Through their agency visible light may have replaced ultra-violet light in activating the mixture and producing greater variety of organic molecules. The great majority of these reactions, like the great majority of photochemical reactions studied today, must have taken place with an overall loss in potential chemical energy. The first crucial step in the evolution of photosynthesis was evidently achieved with the appearance of the magnesium-porphyrin derivative, chlorophyll, which can promote a photochemical reaction which is not followed by retrograde back-reactions, so that potential chemical energy could be accumulated in stable compounds. It will be remembered that chlorophyll by itself is not capable of this and that the evolution of a suitable macromolecular complex to provide the environment in which chlorophyll could manifest its special properties would be an essential part of this step.

The evolution of the photosynthetic system

In present-day plants photosynthesis consists of a number of partial reactions and it is reasonable to suppose that these were brought together one by one. Probably the last to be added was that which enables the plant to use water as a source of hydrogen and to dispose of the oxidizing fragment by the liberation of free oxygen. The photosynthetic bacteria are present-day organisms which seem to have retained a type of photosynthesis which went on before this second crucial development occurred. Their anaerobic nature and dependence on substances such as molecular hydrogen and hydrogen sulphide suggest that they evolved at an early stage when the atmosphere was still without oxygen (that is, *anoxic*). It is significant that this limited type of photosynthesis is only found in these unicellular organisms, which from the relative simplicity of their structure seem to be among the most primitive kinds of organism. Organisms which possess the photochemical system which enables them to liberate oxygen from water and so be independent of a supply of reducing substances are structurally more advanced than the bacteria and evidently evolved at a later date. Several variants of this step, utilizing different accessory pigments, were successful enough to have persisted until modern times in the various algal groups of marine and freshwater habitats. It is indeed conceivable that the pigmentation of the blue-green algae, which is probably the oldest algal group, represents chromatic adaptation to the quality of the weak light penetrating the atmosphere in the anoxic period when it was charged with reddish-coloured hydrocarbons as the atmosphere of Jupiter is today. Some algae are able to revert to the more primitive type of photosynthesis when adapted under anaerobic conditions, but the structurally most advanced plants have lost the ancillary enzyme system enabling them to do this and can only photosynthesize with water as hydrogen donor.

Photosynthesis and the general course of evolution

From various kinds of geochemical evidence it appears that the Earth's atmosphere was anoxic until some time in the Pre-Cambrian era, perhaps 1000 million years ago. There is evidence of life being already in existence during the anoxic period. The Soudan Iron formation of Minnesota, which is dated at 2700 million years, has been found to contain hydrocarbons which, while not certainly of biological origin, may be derived from chlorophyll or related compounds. Fossil organisms, resembling certain present bacteria and algae, have been found in the Gunflint formation of Ontario, dating back to 1600 million years. It seems to have been round about 1000 million years ago that the oxygen-evolving step of photosynthesis appeared. More elaborate organisms, such as protozoa, corals and sponges, seem not to have appeared until just before the Cambrian, which began 600 million years ago, and the increase in abundance and variety of the fossil aquatic fauna which dates from then suggests that dissolved oxygen had become available and that perhaps by this time a low concentration was present in the atmosphere. Anaerobic mechanisms are not efficient in providing energy for life processes and appear incapable of sustaining any very elaborate form of life, for present-day obligate anaerobes are all unicellular. The much more efficient aerobic respiratory mechanisms made possible organisms of greater activity size and complexity and thus undoubtedly influenced evolutionary advance.

The great uniformity in basic biochemical mechanisms found in all organisms from bacteria to higher plants and animals suggests that these mechanisms were perfected early on in the course of evolution, presumably by the end of the Pre-Cambrian. After this, development of the photosynthetic apparatus was largely on the anatomical and morphological levels to meet the change from the aquatic environment to land. Short-wave ultra-violet light is lethal to life and when the concentration of oxygen was below 1 per cent shielding by ozone in the atmosphere would still be so slight that life could only exist under water, which absorbs these wavelengths. With 10 per cent of oxygen in the atmosphere sufficient ozone would be present to reduce the penetration of ultra-violet to a level at which it was possible for life to emerge and colonize the land.

For free-floating aquatic algae supplies of water, carbon dioxide, dissolved mineral salts, and, given a degree of buoyancy, light also, are immediately available to the cell, but large size is a serious disadvantage because it makes for inefficiency in exchange with the environment. Once plant life became attached to a substratum, however, large size was of biological advantage in competition for light. When the attached type of plant colonized the land structural complexity, involving not only special absorbing organs but transporting and mechanical tissues, became necessary in addition as supplies of water and salts and light were spatially separated and support in air was essential. The shortage of water in most terrestrial habitats called for the development of those features of anatomy and physiological behaviour which, as seen in Chapter 2, are necessary to hold the balance between the supply of materials for photosynthesis and the control of the concomitant water loss. For the plant physiologist much of the story of the evolution of the plant kingdom, from alga to flowering plant, is interpretable in terms of the gradual perfection of these arrangements. The anatomical structure of the leaf was stabilized fairly early in the course of evolution of the land plant. Although in the mosses and liverworts all sorts of variants are encountered it is significant that the anatomy of a fern leaf does not differ in general plan from that of a flowering plant. The most recent improvements have been in the development of the shoot system into a more efficient light-diffusing and -absorbing system. The small-leaved tree with its adjustable and expendable leaves and the grass with its vertical short-lived leaves represent the peak of photosynthetic achievement. It is also interesting to note that of the variants in pigment systems found amongst the algal groups only one, that of the green algae, Chlorophyceae, appears in the higher plants. It is perhaps more likely that associated features, rather than a greater intrinsic suitability of this particular photosynthetic system for terrestrial life, are responsible for the overwhelming evolutionary success of this line of plants.

Equilibration to present conditions must have been gradual. The enormous deposits of coal, which are, of course, the remains of once-living plants, suggest that the concentration of carbon dioxide was higher in Carboniferous times, 300 million years ago. This was also probably true of the Cretaceous but it is not clear why extensive coal deposits were not laid down in this period too. Nevertheless it is perhaps significant that present-day plants are adapted to photosynthesize with higher concentrations of this gas than normally occur in the atmosphere. As already indicated the attainment of the present-day concentration of 20·95 per cent by volume of oxygen was probably gradual. It seems that all the oxygen at present in the atmosphere has been produced by the agency of green plants. They seem, indeed, to have produced an environment unfavourable for their own growth; high concentrations of oxygen strongly inhibit photosynthesis and even the concentration normally present in air reduces its rate somewhat.

The present rôle of photosynthesis in the Earth's economy

The importance of photosynthesis in maintaining an atmosphere suitable for terrestrial animal life was realized by the pioneer investigators Priestley and Ingen-Housz. The former, in 1771, carried out a famous experiment in which it was shown that a sprig of mint restored the capacity of air to maintain the life of a mouse after it had been "made noxious by mice breathing and dying in it". In his book *Experiments upon vegetables, discovering their great power of purifying common air in the sunshine and of injuring it in the shade and at night*, published in 1779 (Plate I), Ingen-Housz confirmed and extended Priestley's findings. He rather over-emphasized what we now recognize as the liberation of carbon dioxide by plant respiration, thereby starting the tradition of removing flowers from the sick room at night which still results in a great deal of quite pointless labour. However, Ingen-Housz's assessment of the general situation was perfectly correct; largely through the agency of photosynthesis the oxygen and carbon dioxide contents of the air are maintained at remarkably constant values. From the estimates of the total yield of photosynthesis on the earth which will be discussed below it follows that it takes about 10,000 years to renew all the oxygen in the atmosphere. In recent years there has been an increase in the carbon dioxide concentration, amounting to 0·00007 per cent by volume of air per annum, attributable to the burning of fossil fuels.

The other major ecological function of photosynthesis, that of providing the organic material on which non-photosynthetic life depends for its supply of energy, was made clear, as we saw in the first chapter, by Mayer in 1845. Estimation of the total annual yield of photosynthesis on Earth was attempted by various nineteenth-century workers, mainly on the basis of agricultural returns. Using improved statistical data of this sort and making intelligent guesses about the performance of natural vegetation Schroeder published in 1919 an estimate of $1·63 \times 10^{10}$ metric tons of carbon per annum for fixation by land plants. As we shall see this was quite near the mark, but he had no basis for estimating the contribution of the phytoplankton of the oceans, at first dismissing this as negligible, then later conceding that it might amount to as much as twice that of land plants. With the rapid increase in the human population of the earth it is becoming of practical importance to have a detailed picture of the yield of photosynthesis or, as it is called by ecologists, the *primary productivity*, of the different types of vegetation as a basis for the assessment of the resources available to us and for the intelligent use of land. Such a stocktaking was one of the main objects of the International Biological Programme which was inaugurated in 1966. This included, as well as estimation of the primary productivities of representative types of vegetation, study of the utilization of the products of photosynthesis in various eco-systems and of the photosynthetic performance of particular crop plants under different climatic conditions.

Primary productivity: the land

A reasonably accurate estimate of the annual yield of photosynthesis by annual plants growing in terrestrial habitats may be arrived at by harvesting the plants at the end of the growing season and determining their dry weight, which after suitable corrections to allow for unharvested roots and material lost by grazing and leaf-fall, can be taken as a measure of the total carbon fixed. Applied to perennial plants the method becomes vastly more laborious and complicated but nevertheless can be made to yield reliable results. A correction may be made for respiration to give gross primary production, but for many ecological purposes nett primary production is the more useful value to have.

Some representative results for nett annual production and mean nett photo-synthesis rate in the growing season for terrestrial vegetation are given in the

TABLE

Nett annual product in metric tons of organic matter per hectare (1 metric ton per hectare = 0·4 English tons per acre), and mean growing season productivity, in grams of carbon fixed per day, of various types of vegetation. Data selected from Westlake (*Biol. Rev.*, 1963, **38**, 385).

	Organic matter m.t./ha./year	g C/m²/day
Birch forest, England	8·5	2·2
Pine forest, England	16	2·0
Tropical rain forest	59	7·6
Maize crop, Minnesota	24	8·1
Sugar cane, Java	87	11·0
Lake Erken, Sweden	1·4	0·19
Marine plankton, Isefjord, Denmark	2·4	0·32
Littoral seaweed, Nova Scotia	18·5	2·5
Sublittoral seaweed, Nova Scotia	32	3·9
Marsh, Minnesota	23	6·0

accompanying table. The primary productivity of extreme habitats such as arctic tundra and semi-shrub desert, not given in the table, is of the order of 1 metric ton of organic matter per hectare per year. From such results

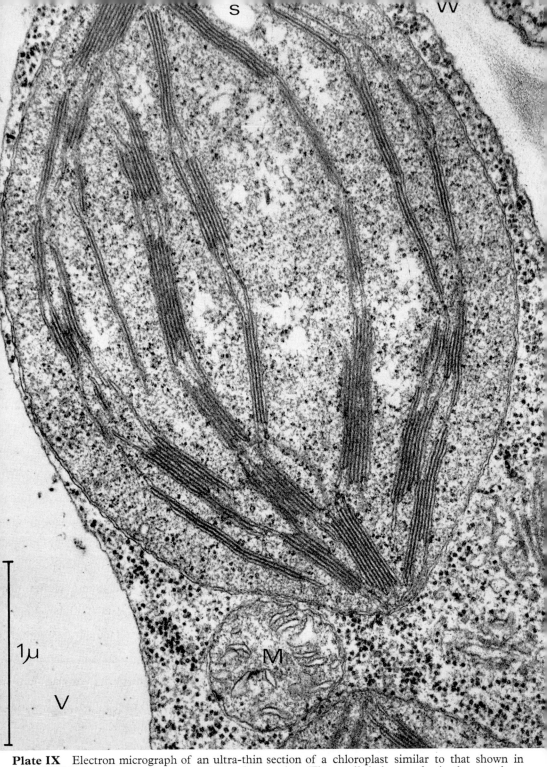

Plate IX Electron micrograph of an ultra-thin section of a chloroplast similar to that shown in Plate VIII but stained to accentuate the nucleoproteins. The small dark granules in the cytoplasm and in the stroma of the chloroplast are ribosomes, which are rich in ribonucleic acid. V, vacuole; W, cell wall; M, mitochondrion; S, starch grain, ×49,000.

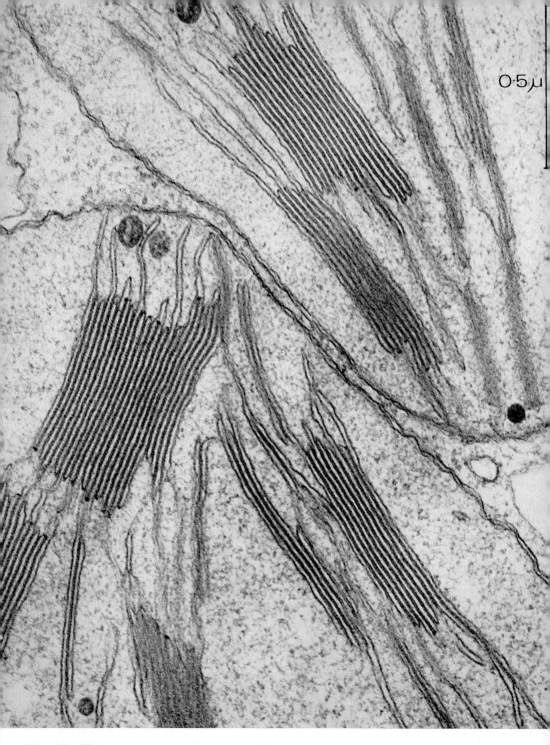

0·5μ

Plate X High-power electron micrograph of an ultra-thin chloroplast section from a leaf of spinach (*Spinacia oleracea*). Parts of two chloroplasts are included in the picture. The nature of the thylakoids and their arrangement in the granum are shown. The dense round bodies are lipid globules. ×88,000.

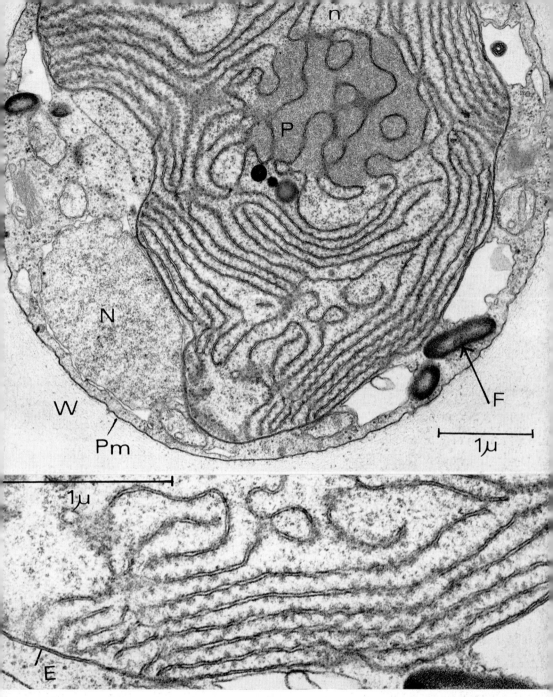

Plate XI (*a*) Electron micrograph of an ultra-thin section of the unicellular red alga, *Porphyridium cruentum*. The photosynthetic lamellae, which consist of single thylakoids, are not arranged in grana as they are in the chloroplast of higher plants and a characteristic structure, the pyrenoid, perhaps concerned with starch synthesis, is present. P, pyrenoid; N, nucleus; W, wall or capsular material; F = floridean starch; n, DNA-containing area of the plastid; Pm, plasma membrane. ×26,000. (*b*) Enlarged portion of (*a*). This shows the special "granules", visible after glutaraldehyde fixation, associated with the photosynthetic lamellae in red algae. These granules probably contain the biliprotein accessory pigment and have not been found in plastids from other plants. E, chloroplast envelope. ×52,000.

Plate XII   Culture of *Chlorella* on the pilot-plant scale at the Microalgae Research Institute of Japan, December 1959.
(*a*) A small pond used for growing inoculum, in the empty condition. The radial arms which distribute carbon dioxide and carry nylon brushes to prevent sedimentation of cells on the bottom are shown.
(*b*) A series of three large culture ponds. In the background are storage tanks and buildings housing the centrifuges and other equipment.

several generalizations can be made. Firstly it will be seen that the mean rate of photosynthesis in the growing season shows less variation than the annual production, that is, ignoring the results for the planktonic communities for the moment, a variation of 5·5- as compared with one of over 60-fold. This implies that the annual production is determined by the length of the growing season rather than by other factors related to latitude. In fact the maximum rate of photosynthesis by plants in the arctic is about the same as that of plants in more favourable climates. The amount of light available to the arctic plants in the summer is more than that received by tropical vegetation in an equivalent period (see p. 14) and the greater productivity of the latter is largely due to the longer period for which they can make use of it. The length of the growing season may be taken as the number of days with a mean temperature of more than 5°C, provided that the water supply is adequate during this period. The importance of water supply is obvious; the highest productivities and rates of photosynthesis are shown by rain forest, marsh and swamp vegetation and irrigated crops, and the productivity of littoral seaweed, which may be desiccated in between tides, is distinctly less than that of sublittoral weed on the same coast which is immersed all the time, although here the situation is complicated by factors in addition to the direct effects of water deficiency. Evergreen forest is more productive than comparable deciduous forest, probably because it is able to take advantage of any conditions favourable for photosynthesis, whereas with deciduous trees there is a lag at the beginning of the growing season while photosynthetically active leaves develop. Lastly it is interesting to note that cultivated crops, in spite of all the fertilizer and cultivation which are lavished on them, are not any more photosynthetically efficient than natural vegetation in corresponding situations, in fact, often distinctly less so.

Primary productivity: lakes and seas

Estimation of the productivity of the suspended plant life of lakes and oceans is not so straightforward as for land plants. Because the growth rate of phytoplankton may be high and there may be equally rapid loss through sedimentation and grazing by zooplankton, the rate of turnover can also be high and then the photosynthesis of phytoplankton bears no simple relationship to the magnitude of the standing crop. Direct determinations of rate of photosynthesis are therefore necessary. Two principal methods for doing this have been used. One, depending on measurement of oxygen production, is only accurate in relatively rich waters. The second, relying on determination of assimilation of radiocarbon supplied as bicarbonate, introduced by the Danish oceanographer Steemann Nielsen in 1952, is a far more sensitive method and is now widely used. Both methods have disadvantages in common.

Determination of oxygen or carbon dioxide in the unconfined waters of the sea or a lake have been used to estimate rates of photosynthesis, but, because of water movements and the difficulty of assessing exchanges with mud and air, do not give very reliable results. For accuracy it is necessary to enclose the water samples in bottles during their period of photosynthesis, but this involves changes in the environment and in the amount of bacterial activity, both of which may affect photosynthesis considerably. The intensity and quality of light varies as it passes through water and, moreover, not only may the density of phytoplankton vary at different depths but the cells may vary in their adaptation to light conditions, as we saw in Chapter 3. Ideally, therefore, determinations should be made with samples taken at different depths and kept suspended *in situ* at these depths during the period of the determination, the total photosynthesis taking place under unit area of the water surface being obtained from the results by integration over the depth of the water-column. This is usually easy in lakes and inshore waters, but it is scarcely practical when it involves the expense of keeping an ocean-going research ship hove-to for twelve or twenty-four hours. Various means of simulating underwater conditions on deck and of estimating primary productivity from factors which can be measured immediately at the sampling station have been devised to avoid this difficulty. To obtain an accurate picture of the total annual production many individual determinations must be made to take account of the horizontal patchiness of phytoplankton and the large day-to-day variations in its activity which take place.

Given sufficient photosynthesis to produce enough oxygen to be measured accurately after a short period of illumination, the oxygen method probably gives a good measure of the total organic matter produced. The radiocarbon method suffers from more uncertainties. The interrelations between respiration and photosynthesis are complex (p. 87), so that it is difficult to know to what extent freshly assimilated carbon is respired and consequently whether the radiocarbon method measures nett or gross photosynthesis. In the original radiocarbon technique the sample was filtered at the end of the exposure and photosynthesis estimated from the radioactivity of the material retained on the filter. Clearly this will only give an accurate estimate if all the products of photosynthesis are retained in the cells, but, as we have seen (p. 83), considerable loss of an early product, glycollic acid, from photosynthesizing cells can take place under certain circumstances. Carbohydrates and other soluble products may also be lost from phytoplankton cells. Most published determinations of primary productivity in aquatic habitats have not taken this possible liberation of extracellular material into account. Both measurement of radiocarbon fixation in dissolved organic products in filtrates and comparison of oxygen (a measure of total photosynthesis) and radiocarbon determinations with natural phytoplankton populations suggest that on the average these values should be increased by 25 per cent to allow for such loss from cells. As extracellular products of phytoplankton photosynthesis are probably assimilated by bacteria, they could indirectly provide an important food source for zooplankton.

The values for phytoplankton communities given in the table are for waters of moderately high productivity. Data are really too sparse to put forward firm values but perhaps the most productive waters, such as the Benguela current off SW Africa in which upwelling brings rich supplies of plant nutrients to the surface, have an annual production of the order of 15 metric tons of organic matter per hectare per year. The waters of lakes in rocky mountainous areas and most of the oceans are poor in plant nutrients, and have productivities perhaps of 0·5 metric tons organic matter per hectare per year or less. The principal factor determining the productivity is the concentration of plant nutrients in the water. In nutrient-poor waters, however, light is able to penetrate to greater depths (see p. 32) so that the phytoplankton has a greater volume of water to draw on and nutrient limitation is thus offset to a certain extent. Turbulence has an important effect, since if it carries cells up and down between the surface and the depths the average light intensity reaching the population is reduced. In fact it seems that the chief factor limiting phytoplankton growth in unfrozen seas and lakes in the winter is turbulence, and growth begins as soon as increasing light and calmer weather in the spring bring the average amount of light received by the cells above the compensation point.

It is clear that even the most productive natural waters make a poor showing beside terrestrial vegetation. There are several good reasons why this should be so. As we saw in Chapter 3, the extent to which phytoplankton can adapt itself to make efficient use of light in photosynthesis is limited and that near the surface is exposed on bright days to intensities which are actually inhibitory. Furthermore, much of the available light may be absorbed by coloured dissolved materials, suspended detritus and the water itself. On the other hand the higher plant is an excellent light-diffusing system in which unduly high intensities on the chloroplasts in the upper leaves are avoided and much light is allowed to penetrate lower down. Furthermore there is usually little except green tissue above soil level to intercept the light. The maximum amount of chlorophyll above a square metre of soil surface in evergreen rain forest is about 13·3 g, as compared with a maximum of about 1·0 g in the illuminated part of the water-column below a square metre of surface. In addition it is only the nutrient salts in the immediate vicinity of phytoplankton cells which are accessible to them, with the result that the supply easily becomes limiting for growth and thus for photosynthesis. The higher plant, having roots which can extend through soil and absorb supplies of salts at a distance, is again at an advantage.

The total yield of photosynthesis on Earth

On the basis of the more recent estimates Schroeder's value for the total net productivity of the land area of the Earth must be revised upwards, and Vallentyne (1966) gives it as 2·2 to 3·2 $\times$ 10^{10} metric tons of carbon per year, an average of 1·4 to 2·1 tons of carbon per hectare per year.

From the available data and multiplying by a factor of 1·25 to allow for loss of extracellular products of photosynthesis, Vallentyne estimates that the fixation of carbon in the oceans averages 0·6 to 0·8 metric tons of carbon per hectare a year, less than a half that of the land but, because of their greater extent, amounting to a total, 2·2 to 2·8 $\times$ 10^{10} metric tons of carbon per annum, which is about the same as that for the land area. In terms of conversion of the energy of visible light into the potential chemical energy of organic matter the efficiency is low, about 0·2 per cent for the oceans, 0·4 per cent for the land and between 0·2 and 0·3 per cent for the earth as a whole. Nevertheless the total yield of photosynthesis, something of the order of 5 $\times$ 10^{10} metric tons of carbon or 12·5 $\times$ 10^{10} metric tons of organic matter per annum is enormous by human standards. The amount of food consumed by the earth's human population is only about 1/200th of this. If a year's yield of photosynthesis were amassed in the form of cane sugar it would form a heap over 2 miles (3·2 km) high and with a base of 43 sq miles (111 km²).

The exploitation of photosynthesis

Fortunately for agricultural man the photosynthetic mechanism is self-regulating to a remarkable degree and no understanding of the way in which it works is necessary to exploit it. Moreover the supply of light from the sun is so copious and unfailing that it has not so far been necessary to worry if only low efficiencies are attained in using it. Perhaps the end of this era is in sight. As we have just seen, the average efficiency of photosynthesis on Earth is much below the maximum efficiencies of about 25 per cent which can be attained in the laboratory. Even sugar cane, the most efficient in photosynthesis of Man's crops, only achieves an average of 1·4 per cent over the year. When the fact that the consumable part of the crop is only a fraction of the total dry matter produced by the plant—for maize it is about 24 per cent—then the efficiency in terms of utilization of solar radiation of conventional agriculture is seen to be poor indeed. The principal reason for this is that most crop plants are annual, so that for some weeks after sowing most of the light falls on bare ground and later the leaf canopy only reaches maximum photosynthetic efficiency for a week or so in the middle of the growing season (Fig. 40).

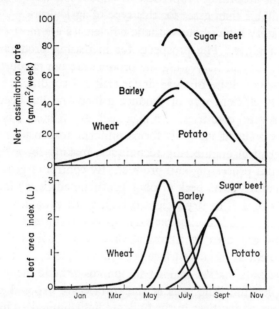

Figure 40 Smoothed curves showing the average variations with season in rate of photosynthesis (nett assimilation rate, *E*, and leaf area index, *L*, of various crop plants grown at Rothamsted Experimental Station. (Redrawn from D. J. Watson (1956) in *The Growth of Leaves*, p.180; Butterworths.)

Improvements in the photosynthetic efficiency of crops are clearly desirable, especially in densely populated countries. Much, of course, can be accomplished by improvements in irrigation, fertilization and husbandry and by breeding more efficient varieties. Better yields could also be obtained by increasing the concentration of carbon dioxide available to the crop, a matter which was discussed on page 20. However, we may also consider some less conventional methods. Pirie has suggested that it might be a good plan to reverse the usual procedure and, instead of deciding what products are needed and then growing crops to produce them, to select a plant which gives a high yield under the local conditions and then find uses for the material which it produces. In pursuance of this idea Pirie has devised methods for extracting high-quality protein from leaves on an economic scale. To reduce photosynthetic inefficiencies to a minimum it would be best to arrange things on a continuous basis; that is to say, have a crop plant with complete ground cover all the time which produces leaves continuously during the growing season, the crop being taken in the form of those leaves which are past their photosynthetic prime. A grass lawn approximates to this sort of system but has rather poor light-absorbing properties. It should not be too difficult to find plants more suitable than grass for this type of agriculture.

In the laboratory high photosynthetic efficiencies are most easily obtained with unicellular algae. The prospects for increasing production of phytoplankton in the sea and harvesting the product are not encouraging, but the possibility of mass culture of algae under artificial conditions has some promise as a highly efficient way of producing food and has been investigated intensively in several countries. An alga such as *Chlorella* has many advantages from this point of view. It forms uniform suspensions which can be handled by chemical engineering techniques, making for a high degree of standardization in processing and product. By control of growth conditions it can be maintained at its peak of photosynthetic efficiency indefinitely. As much as 60 per cent of its dry weight is high-grade protein and it contains a minimum of nutritionally useless material. In dense cultures increase becomes proportional to the area of illumination and so mass cultures must be arranged to present as large a surface per unit volume to the light as possible. There is the same situation making for photosynthetic inefficiency as with natural phytoplankton—the cells at the surface are exposed to light of inhibitory intensity while those in the bulk are light-limited. This may be partially overcome by high rates of stirring so that the cells are in effect exposed to flashing light (see p. 5). A more practical device is to use plastic or glass diffusing cones immersed in the culture with their bases presented to the light, so that this is supplied over a greater surface at a lower intensity. By means of such cones yields of cultures exposed to intensities equivalent to full sunlight have been doubled.

Mass cultures on a pilot-plant scale have now been tested in several countries. Perhaps the most successful to date is that of the Microalgae Research Institute of Japan just outside Tokyo (Plate XIIb). Here *Chlorella* is grown in shallow circular ponds having a total area of 1 acre (0·4 hectare). The cells are kept in suspension and carbon dioxide is distributed by rotating radial arms (Plate XIIa). The culture is continuous, the algal suspension being circulated through centrifuges to harvest the alga and the medium returned to the ponds after recharging with nutrient salts. These ponds have actually yielded 32 metric tons of dry algal material per hectare per year, the average daily yield being about 4 g carbon per metre². By reference back to the table on page 100 it will be seen that this is a reasonably good rate of production but still well under that which sugar cane at its best can attain. However, considered in terms of protein production per acre these *Chlorella* cultures greatly surpass any other crop, yielding 23 times as much protein as a grass crop, 33 times as much as a bean crop, and 260 times as much as would be obtained in meat from grazing cattle. The algal material produced has high nutritive value and can be incorporated quite palatably in soup, ice-cream and other dishes. Nevertheless the high cost of this cultivated *Chlorella* precludes its production on an economic scale. The capital cost of the ponds and ancillary machinery is high and the carbon dioxide, produced by burning oil, and the harvesting

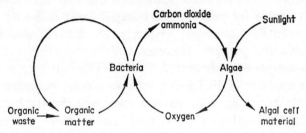

Figure 41 Diagram showing the interrelations of algal and bacterial activities in sewage lagoons.

by high-speed centrifuge are also expensive. No doubt with further improvements in technique these costs could be brought down but it does not seem likely that mass-culture of algae will be an economic way of producing food in the immediate future.

A more practical possibility is to combine algal culture with sewage disposal. If domestic sewage or industrial waste water, such as that from canning factories, containing much organic matter, is exposed to the light in shallow lagoons, or oxidation ponds, associated growth of algae and bacteria takes place and results in rapid stabilization of the sewage. The decomposition of the organic matter by the bacteria produces carbon dioxide and ammonia, both of which stimulate algal photosynthesis while this in turn produces oxygen which promotes the bacterial breakdown of the organic matter (Fig. 41). In addition, as we saw on page 78, the algae may use certain organic substances directly by photoassimilation and there is some evidence that they may produce antibiotic substances active against pathogenic bacteria—certainly such organisms disappear rather quickly in oxidation ponds. This biological system is thus capable of rapid and efficient conversion of noxious organic matter into the innocuous material of algal cells. It operates naturally in many situations; if it did not, epidemics of water-borne disease would doubtless be more common than they are in tropical countries. In California extensive studies of the basic principles of the necessary engineering design have been carried out and photosynthetic sewage disposal is now practised in many regions where sunshine can be relied upon and land is not prohibitively expensive. The yield of algal material may be as much as 1·5 tons dry weight per million gallons of sewage or 150 tons per hectare per year, twice as much as the highest yields from terrestrial vegetation. This material is itself liable to give rise to pollution problems if discharged into water-courses in which it is unable to grow, but in any case such a waste of potentially valuable organic matter should not be contemplated. As the predominant algae in oxidation ponds are species of *Chlorella* and *Scenedesmus* that have been shown to have good nutritive value for poultry, this material could indirectly serve as a source of human food although, obviously, the material would not itself be acceptable for this purpose. However, a really cheap means of harvesting the algae has not yet been developed, although sand filtration is reasonably effective. As the folly of discharging valuable nitrate, phosphate and other plant nutrients in sewage effluents into lakes and the sea becomes more generally recognized and the need for food becomes more pressing it can be expected that simultaneous sewage disposal and algal culture will inevitably be greatly improved and extensively used.

Photosynthesis and space research

The problems discussed in the latter part of this chapter are just as vital in the microcosms with which Man may colonize space, as for life on Earth as a whole. So far, the atmosphere in space capsules has been maintained in a state fit for breathing by expenditure of stored materials and food has been Earth-grown. Probably this will always be the practical method except on prolonged voyages on which, clearly, a regenerative system would be highly desirable. Both Americans and Russians have investigated the possibility of using algal photosynthesis as a means of balancing the respiration of astronauts. Priestley's experiment with the mouse has been repeated, in a more elaborate way, by Myers in Texas. The mouse was maintained for as long as twenty-four days in a closed system, carefully checked for leaks, containing an illuminated suspension of *Chlorella*. The major difficulty lay in obtaining an exact match in the ratios of exchange of carbon dioxide and oxygen by the two organisms. In Myers' experiment 98 per cent of the desired perfect match was obtained. To meet the requirement of 600 litres of oxygen per day per man an algal exchanger would take up about 720 litres of carbon dioxide, produce about 600 g of dry algal material and require a minimum of about 800 watts of visible light. Because algal photosynthesis is most efficient at low intensities the attainment of this in a form which could be accommodated in a space vehicle involves formidable problems of engineering.

In a semi-permanent space station such as a lunar base the men should as far as possible grow their own food and recycle waste products of their own metabolism to conserve essential elements. Compactness would not be an overriding consideration here and higher plants might present advantages over algae. As Pirie has pointed out, our conventional crop plants would probably not produce fruits and tubers if grown with fourteen-day periods of light followed by fourteen-day periods of dark such as they would get on the moon, and it would be necessary to rely on the vegetative parts, perhaps by extracting leaf protein. Difficulties would doubtless be encountered in obtaining efficient photosynthesis in atmospheres, such as 5 per cent carbon dioxide and 95 per cent oxygen at a pressure of a fifth of atmospheric, that it might be necessary to use, and also in finding species which will tolerate the high salt concentrations present in recycled urine, but these should not be insuperable.

In space or on Earth the sun will always be our major source of energy. The solar radiation reaching the Earth's surface amounts to 5×10^{23} calories per annum or as much in three days as could be produced by burning the world's entire reserves of coal, petroleum, natural gas and tar together with all the forests. Photosynthesis is at present the best means that we have of making use of this colossal source of power and with greater understanding of plant physiology and biochemistry and with developments in bio-engineering techniques our use will no doubt be made a great deal more effective. That Man will ever devise an artificial system which will approach, let alone surpass, the effectiveness of the plant as a photosynthetic system seems extremely doubtful.

Further Reading

Calvin, M. (1965) Chemical evolution. *Proc. Roy. Soc.* A, **288**, 441–66.

Echlin, P. (1966) Origins of photosynthesis. *Science Journal*, April 1966, 7 pp.

Eley, J. H., Jr., and Myers, J. (1964) Study of a photosynthetic gas exchanger. A quantitative repetition of the Priestley experiment. *Texas Journal of Science*, **16**, 296–333.

Evans, L. T. (1968) Photosynthesis under natural conditions. *Penguin Science Survey 1969: Biology*, edited by A. Allison, pp. 73–92. Penguin Books, Harmondsworth.

Fogg, G. E. (1971) Recycling through algae. *Proc. Roy. Soc.* B, *179*, 201–207.

Gaffron, H. (1962) On dating stages in photochemical evolution. In *Horizons in Biochemistry*, pp. 59–89. Academic Press, New York.

Goldman, C. R. (1966) Primary productivity in aquatic environments. *Memorie dell Instituto Italiano di Idrobiologia*, 18 suppl. (1965). University of California Press.

Microalgae Research Institute of Japan, Reports (1961) Vol. II (1). Japan Nutrition Association, Tokyo.

Myers, J. (1964) Use of algae for support of the human in space. In *Life Sciences and Space Research* II edited by M. Florkin and A. Dollfus, pp. 323–36. North Holland Publishing Co., Amsterdam.

Oswald, W. J., and Golueke, C. G. (1968) Harvesting and processing of waste-grown microalgae. *Algae, Man, and the Environment*, edited by D. F. Jackson, pp. 371–89. Syracuse University Press.

Pirie, N. W. (1966) Towards the seleno-microcosm. *New Scientist*, 2 June 1966, 574–6.

Prediction and Measurement of Photosynthetic Productivity. Proceedings of the IBP/PP Technical Meeting, Trebon, 14–21 September 1969. Centre for Agricultural Publishing and Documentation, Wageningen, 1970.

Steemann Nielsen, E. (1963) Productivity, definition and measurement. In *The Seas* edited by M. N. Hill, vol. 2, pp. 129–64. Interscience, New York and London.

Tamiya, H. (1957) Mass culture of algae. *Ann. Rev. Plant Physiol.*, 8, 309–34.

Westlake, D. F. (1963) Comparisons of plant productivity. *Biol. Rev.*, 38, 385–425.

Index

Absorption spectra, 24, **25** (figs. 7 and 8), **28** (fig. 9), 29, **30** (fig. 10), 37, **38** (fig. 15), **45** (fig. 18), 49, **50** (fig. 19), 56, **57** (fig. 26)

Accessory pigments, 4, 29, 30, 36, 46, 53

Acetate, 73, 74, 79, 80, 83, 95

Acetyl coenzyme A, 70, 83

Action spectra, 29, **30** (fig. 10), **45** (fig. 18)

Adaptation to carbon dioxide concentration, 18, 20
— to light intensity, 34, 35, 102

Adiantum, 39

Ageing, 52, 91

Alanine, 62, 82, 83

Aldolase, 66

Algae, anaerobically adapted, 7, 59, 72, 75, 92
— compared with higher plants, 37, 41, 68, 96-8, 103-4
— mass culture, 106-9, **Plate XII**

Amino-acids, 62, 68, 82, 83, 94

Ammonia, 75-8, 82, 83, 94, 107

Anabaena, 76, 78, 89

Anacystis, 47, 90

Ankistrodesmus, 76

Anthocyanins, 41

Arnold, 6, 45

Arnon, 10, 11, 52, 59, 69, 74

Ascorbic acid, 52, 54, **58** (fig. 27)

Aspartic acid, 68, 82, 83

ATP (adenosine triphosphate), 8-11, 42, 51-4, 58, 59, 64, 65, 67, 69, 70-80, 83, 84, 86, 88, 90

Bacteria, green sulphur, 6, 7
— indicators of oxygen, 2, 36, **Plate IIa**
— photosynthetic, 6-8, 10, 27, 47, 49, 54, 57, 59, 69, 72, 75, 77, 78, 96
— purple non-sulphur, 7, 26, 50, 51, 69
— purple sulphur, 6, 7, 54
— relations with algae, 103, **107** (fig. 41)

Bacteriochlorophyll, 6, **25** (fig. 7), 26, 32, 50

Baeyer, 3

Barley, 68, **105** (fig. 40)

Beech, 39, 40
— copper, 41

Benson, 8, 10, 11, 64

Bicarbonate ion, 14-16, 18

Biliproteins, 24, 26, 27, 47

Bishop, 56

Blackman, 4

Blinks, 46, 47

Blue-green algae, 24, 27, 29, **30** (fig. 10), 37, 46, 47, 56, 68, 76-8, 89, 96

Brown, 18

Calcium carbonate precipitation, 16, 74